UXO Team Leadership:
How a leader creates and handles an effective unexploded ordnance (UXO) team

by
George J. DeMetropolis

ISBN: 1-58112-187-3

DISSERTATION.COM

Parkland, FL • USA • 2003

UXO Team Leadership:
How a leader creates and handles an effective unexploded ordnance (UXO) team

Dissertation.com
USA • 2003

ISBN: 1-58112-187-3
www.Dissertation.com/library/1121873a.htm

UXO TEAM LEADERSHIP:

HOW A LEADER CREATES AND HANDLES AN EFFECTIVE

UNEXPLODED ORDNANCE (UXO) TEAM

A Final Dissertation

Presented to the

Faculty of the

School of Business Administration

Kennedy-Western University

In Partial Fulfillment

of the Requirements for the Degree of

Doctor of Philosophy in

Business Administration

by

George DeMetropolis

Kahului, Hawaii

Abstract of the Study

UXO Team Leadership:

How A Leader Creates and Handles An Effective

Unexploded Ordnance (UXO) Team

by

George DeMetropolis

Kennedy-Western University

THE PROBLEM

The primary purpose of this study was to provide an analysis of the effectiveness of leadership through the interpersonal relationship between a team leader and a team. This research paper attempted to define "how" leaders create and handle effective teams. Specifically, this study was focused on UXO teams in a UXO environment performing a UXO clearance project still in progress. It was the aim of this study to provide information that will be beneficial to team leaders and will contribute to improvement of UXO team leadership techniques.

METHOD

This research evaluated the relationship between the leadership demonstrated by team leaders and the effectiveness and productivity of the teams. Adoption and use of Frank LaFasto and Carl Larson s (2001) six key dimensions (focus on the goal, ensure a collaborative climate, build confidence, demonstrate sufficient technical know-how, set priorities, and manage performance) provided the basis for this study. Frank LaFasto, Ph.D. and Carl Larson, Ph.D. developed the survey instrument adopted for this study, known as the Collaborative Team Leader Instrument, from an evaluation of approximately 600 team leaders and measures team leader effectiveness across the six key dimensions.

FINDINGS

The findings identified in this study provide some empirical support and insight into the strengths and weaknesses found in one case study. The findings revealed positive correlation in many of the researched areas within each hypothesis in the relationships and perceptions between team leaders and team members. However, the findings also indicated considerable negative correlation within one of the dimensions, the demonstration of sufficient technical know-how, which was actually expected to be one of the strongest.

Overall it is concluded that this research has made an important contribution towards defining: how a leader creates and handles an effective unexploded ordnance team. The findings of this study should be beneficial to team leaders and contribute to improvement of UXO team leadership techniques.

TABLE OF CONTENTS

CHAPTER

2. A REVIEW OF RELATED LITERATURE

(continued)

CHAPTER

4.   DATA ANALYSIS

(continued)

CHAPTER

5.   SUMMARY, CONCLUSIONS, AND

CHAPTER

5. SUMMARY, CONCLUSIONS, AND

RECOMMENDATIONS

(continued)

FIGURE

LIST OF TABLES

TABLE

TABLE (continued)

Chapter 1

INTRODUCTION TO THE STUDY

The study of leadership is a widely documented subject. Many people today from various levels within an organization are seeking to understand the concepts of leadership. As organizations are reorganizing due to globalization and the need to be innovative in a constantly changing environment, the value of sound leadership skills, continuous performance improvement, and effectiveness are becoming more and more important at the team level. Csoka (1998) believes that only human creativity and commitment   not technology   can deliver success through sound leadership practices (p. 7).

The use of teams in the work environment has expanded within various industries, but is not new to the Unexploded Ordnance (UXO) industry involved in UXO clearance. Specifically, these teams are involved in UXO clearance projects throughout the nation and world. Typically, the work site is geographically separated from the corporate office by many miles. These work teams, composed of non-UXO qualified personnel and UXO qualified personnel, or just UXO qualified personnel all led by a UXO qualified team leader, perform the core work of the company. They convert technical knowledge and labor into clearance services for a client. Cohen and Bailey (1997) reflect the traditional model for a work team as being led by a supervisor who directs what needs to be accomplished, how it is to be accomplished, and who will perform the task. The leadership of the work teams being addressed in this study is consistent with this traditional model.

Most of the UXO qualified personnel are former military Explosive Ordnance Disposal technicians and are graduates of the Naval School Explosive Ordnance Disposal. This school varies in length from 42 weeks for Army, Air Force and Marine Corps candidates to 51 weeks for Navy candidates. Navy candidates receive an additional nine weeks of dive school training at the Naval Diving and Salvage Training Center. Explosive ordnance disposal training common to all the services includes the following subject areas:

Core classes: ordnance identification, safety precautions, publications and reconnaissance techniques

Demolition of explosives

Explosive ordnance disposal tools and methods

Biological/chemical ordnance

Ground ordnance: landmines, grenades, booby traps and projectiles, mortars and rockets

Air ordnance: bombs, missiles, gun systems, and aircraft explosive hazards

Improvised explosive devices: homemade bombs and terrorist devices

Nuclear weapons: basic nuclear physics, radiation monitoring and decontamination procedures

Additional training provided to Navy candidates includes:

Underwater ordnance: torpedoes, mines and underwater explosive devices

Underwater tools and techniques: specialized diving and recovery techniques associated with underwater ordnance

Recently Texas A & M University established a civilian course of instruction that duplicates the military curriculum and will qualify UXO personnel at the apprentice level. After fulfilling time and on the job training requirements, these civilian-trained personnel will be promoted to the technician level. It is anticipated that this additional source of qualifying UXO personnel will alleviate the current shortfall within the industry due to the limited number (approximately 600) UXO technicians worldwide.

The targets of this research are UXO work teams involved in a UXO clearance effort of an isolated island that was used as a training range. It was fired upon from the sea, land, and air for 45 years. The island is to be cleared of unexploded ordnance and environmentally restored.

Unexploded ordnance poses an extreme risk to human life and the environment. The UXO detection and clearance methods form the core of the UXO clearance process. These two methods are separate, but interrelated to complete the clearance process. The technical steps of the UXO clearance process are identified in Figure 1 UXO Clearance Process. This process incorporates the various regulatory, technical, and contractual requirements into a single, integrated process. All phases of the process are accomplished in accordance with approved work plans and standard operating procedures. The scope of the work effort may entail surface clearance only or both surface and subsurface clearance. The process starts with the pre-investigation archival search phase during which all existing published documentation in all functional areas is reviewed. Remediation of UXO contaminated areas consists of the following phases:

1. Establishment of work area boundaries and grid map unit work areas
2. Discovery and recording of UXO/OE (other explosive) concentrations, historic property surveys, and environmental conditions report
3. Area preparation and surface sweep
4. Subsurface geophysical detection

5. Excavation of anomalies

6. Clearance of all safe-to-move UXO, UXO-related remnants, target materials, and non-UXO-related materials from the designated work areas

7. Debris and remnant management

8. UXO disposal

9. Quality control

10. Quality assurance.

For surface or Tier I clearance, phases 1-3 and 6-10 are required; for subsurface or Tier II clearance, all the aforementioned phases are completed. Each one of these phases will be described briefly to establish an understanding of the purpose of each phase within the clearance effort and the potential impact of each phase on a team's effectiveness and success.

The establishment of the work area boundaries and grid map unit work areas phase involves the smallest team element within the clearance effort, two people - a surveyor and a UXO escort. The UXO escort guides the survey team, while identifying, marking, and avoiding encountered UXO. The surveyor utilizes a Global Positioning System instrument to establish grid map units that are typically 100 meters square. After the UXO escort determines the location to be clear of any surface UXO or subsurface anomalies, the corners of the grid map unit are determined and marked with wooden stakes. The wooden stake is marked with an eight-digit number that represents the east/west and north/south coordinate values. Establishment of a grid map unit establishes a specific location on the island and facilitates command and control of teams dispersed throughout the island by Range Control.

The discovery phase consists of two activities: area assessment and data recording. Area assessment is a gross characterization of a field condition assessment completed by three personnel providing four functional area perspectives: UXO, historic properties, natural resources, and environment. The UXO supervisor directs the assessment of the grid map unit. The objective for the UXO supervisor is to identify, mark, and record surface UXO and other potential explosive hazards. The types and density of UXO, UXO-related remnants, target materials, and non-UXO related materials within the grid map unit are recorded.

The historic properties person identifies and marks previously recorded and new historic properties within the grid map unit. This information will enable the historic properties functional manager to determine impact and mitigation measures for the protection of historic properties from further clearance efforts.

One representative handles both the natural resources and environmental functional areas. The existing vegetation is documented by type, species, and density. Endangered species are marked and recorded. Additionally, the environmental conditions within the grid map unit, including soil conditions, slope, and terrain are noted.

The next phase is the data recording activity of the assessment phase, which involves further documentation of known historic properties and the initial recording of new sites. Site descriptions, maps, and records for each historic property are prepared. A data recording team consists of one archaeologist and one UXO escort.

Next is the area preparation and surface sweep phase. Activities include the manual removal or controlled burning of vegetation that impedes the ability of personnel to effectively sweep the surface of the ground with detection equipment. Then surface sweep operations are conducted, which entails the removal or clearance of UXO, UXO-remnants, non-UXO related material, and target materials from the surface of the island. Usually the same team, known as the area preparation team, consisting of one UXO supervisor and seven laborers, is responsible for conducting both the area preparation phase and surface sweep phase. For the heavily contaminated areas, a specialized range clearance team is formed and equipped with heavy equipment to better tackle these areas.

One UXO supervisor, five UXO specialists, ten laborers, and one heavy equipment operator for a total of 17 people comprise this team.

A subsurface geophysical detection operation involves detecting and marking subsurface anomalies with a geophysical detection instrument that provides its operator visual and audible signals. The development of new, more sensitive instrumentation has allowed the discovery of deeper and smaller targets. Unfortunately, this has led to an associated increase in project cost and level of effort as each object discovered must be excavated. This team consists of two personnel: a geophysicist and a geophysicist's assistant.

Excavations are conducted to investigate and identify subsurface target anomalies to a contract depth. For this project, the depth is four feet. With the improved detection technology, the result has been an increase in the number of UXO excavated and a dramatic increase in the amount of non-UXO excavated. The manning of an excavation team varies based on the tasking. Typically, it consists of one UXO team leader, five UXO specialists, and one UXO heavy equipment operator.

The next two phases involve the clearance of all safe-to-move UXO, UXO-related remnants, target materials, and non-UXO-related materials from the designated work areas and debris and remnant management. These material management operations involve the collection, inspection, segregation, accounting, demilitarization, and thermal treatment of materials collected from the UXO clearance activities. Material management teams vary in size based on their tasking. Material management teams, responsible for collection of materials, consist of one UXO team leader, one UXO specialist, and one material management specialist. The material management team responsible for manning the explosive inspection point, where materials are inspected, segregated, and accounted, consists of one UXO team leader, two UXO specialists, four material management specialists, and two heavy equipment operators. The material management team responsible for demilitarization and thermal treatment consists of one material management specialist supervisor and four material management specialists.

The UXO disposal phase consists of two activities: UXO access and identification and destruction. Each UXO item is positively identified during the access and identification phase and a specific disposition made as to the most appropriate destruction method. In some instances UXO items can be safely moved and consolidated in a central location; other times the fuzing, size of the item, or location precludes transport, and the item is detonated where it is found.

Contractor quality control inspections and surveillances followed by client quality assurance activities validate quality of work and compliance with the contract. Defects and re-works are managed through the clearance control process by establishing points of control, implementing measurements, regulating the process through feedback, and performing corrective action. The quality control team varies in size from three to five people each led by a UXO team leader. A quality assurance team is not a contractor team and belongs to the client. It will not be addressed in this study.

The performance measure for each phase of the UXO detection and clearance process is depicted in Table 1 below. These measures are specified further in the quality assurance project plan, which elaborates the periodicity of the checks. Immediate feedback to the UXO detection and clearance process enables further refinement to the phases.

Table 1 Performance Measures for Processes

Process	Activity	Performance Measure
Area pre-investigation	Documentation research	Accuracy
Establish grid map unit and work area boundaries	Survey stakeout	Safety, accuracy, speed Production goal
Area Assessment	Field surveillance	Safety, accuracy, speed Production goal
Assessment Review	Quality checks	Accuracy
Area Preparation	Brush removal and selective pruning of trees	Safety, speed Production goal

Process	Activity	Performance Measure
Surface Sweep	Surface UXO clearance	Safety, accuracy, speed No missed UXO items All large pieces of metal removed per contract specification Production goal
Subsurface Detection	Geophysical detector sweep and marking location of suspected UXO	Safety, accuracy in marking locations, speed Complete ground coverage No missed UXO items Production goal
Anomaly Review	Quality checks	Accuracy
Excavation	Excavation of UXO	Safety, accuracy, speed Locate and excavate all marked items Production goal
UXO Identification and Assessment	Identification and disposition assessment of all UXO found	Safety, accuracy, speed Positive identification Accurate disposition assessment
Disposition Review	Quality checks	Accuracy
UXO Destruction	Destruction of UXO	Complete destruction Safety, accuracy
Debris Remnant Management	Collect and sort all debris from UXO	Safety, accuracy, speed No UXO passes through sorting process
Survey/marking	Final land survey	Safety, accuracy Production goal
Records	Documentation of all activities	Accuracy
Certification	Quality checks	Accuracy

Statement of the Problem

UXO clearance projects are extremely costly. One of the major reasons is that there is currently no safe and cheap method of clearing ranges and certifying that they are 100 percent free of UXO contamination. Additionally, despite improvements in detection equipment technologies over the years, the equipment is still not totally reliable. Congress is gradually increasing the budget for improving detection equipment technologies. In the meantime, many UXO clearance projects are being conducted around the world.

In order to distribute the limited funding for UXO clearance and optimize the clearance effort, UXO teams need to be managed and led to maximize production. Thus, the aim of this study is to identify how an effective UXO team leader creates and handles a team to maximize productivity.

Purpose of the Study

The topic of leadership in the workplace is a well-documented subject. However, capturing the reasons one team is deemed to be efficient and another team inefficient is not as well researched and documented. Moreover, an analysis of the most effective leadership style needed in order to create, motivate and manage a UXO team had not been addressed.

This study examined UXO teams involved in a UXO clearance project currently in progress through November, 2003. Some of the teams were determined to be effective and others not as effective based on productivity statistics. The anticipated outcome of this study was that leadership is directly correlated to production levels and team effectiveness. Further, the study attempted to identify those leadership skills that are important for UXO team leaders to use in order to maximize production. This study is based on observations and data collected during a specific case study of an actual UXO clearance project.

The following hypotheses were tested in this research through the use of two survey questionnaires developed by Frank LaFasto and Carl Larson (2001, p.151-154):

1. Clearly defining the team goal contributes to team effectiveness.
2. A collaborative climate increases productivity and team effectiveness.
3. Confidence of team members enhances performance, which results in increased productivity and effectiveness.
4. Job knowledge contributes to improved performance.
5. Establishment of priorities helps achievement of goals.

6. Clear and recognized performance objectives enhance focus and
   result in better performance.

Importance of the Study

In an effort to stay within costs and on schedule, effectiveness of a
UXO team involved in a UXO clearance project is important. From the
perspective of the client, reduced costs and increased production are
desirable outcomes. A contractor that is able to keep the client happy and
the quality of the work effort high will reap the reward of a good reputation
and probably additional contracts. New and prospective team leaders
need to display good leadership in order to increase effectiveness and
efficiency of UXO teams.

An increased level of understanding of worker motivation and
communication requirements should be enhanced through better rapport
and leadership with team members. Attainment of mutual objectives
should be a win-win situation for all. Attitudes, responsiveness, and job
satisfaction should be positive and stimulating for members to continue to
be productive.

The cause-effect relationship between good leadership and
motivation will result in increased production. The establishment of

consistent, good leadership standards should foster productive behavior that result in a competitive advantage.

Scope of the Study

This study is unique because it focuses on the "how" UXO team leaders create and handle effective teams within an actual UXO clearance project. This project is also unique because to this date in the United States it establishes numerous firsts as the largest and longest single UXO clearance effort, largest civilian passenger helicopter operation, and largest chainsaw operation. Added to these records is a large multi-million dollar project on a remote, uninhabited island with very little infrastructure that involves many functional disciplines in a clearance work effort. And of course, the project has to be completed within a legislated schedule and authorized budget. Consequently, new methodologies, techniques, and procedures have been established through the course of this undertaking that will probably be used on future projects. Additionally, lessons learned have been gleaned that will enable others to have a head start without the pitfalls.

Many variables were involved in this study. The variations in the environmental conditions encountered by the work teams as they proceed through the various phases of UXO clearance will affect production.

Previous leadership instruction and experiences of the team leaders will be a factor as well. Personalities, cultures, and individual differences of the various team personnel will come into play. Level of involvement of the leaders with their teams will impact attitudes, motivations, and work behaviors. Diversity in the nature of the tasks will be a challenge.

Some of the limitations of this study will also impact its conclusions. Studying only one, very unique project presents a item as to the validity of the comparison of the results to similar projects. The size of the sample used in the research was rather small and may not be statistically significant. And lastly, the development of surveys as a measurement tool and interpretation of outcomes was subject to biases by the interpreter.

Rationale of the Study

The general significance of this study was that it provided an analysis of the effectiveness of leadership through the interpersonal relationship between a team leader and a team. This research paper attempted to define "how" leaders create and handle effective teams.

Specifically, this study was focused on a UXO team in a UXO environment performing a UXO clearance project still in progress. It was the aim of this study to provide information that will be beneficial to team leaders and will contribute to improvement of UXO team leadership techniques.

Overview of the Study

This study addressed a subject in the field of leadership that needs more research to define and substantiate. Addressing the creation and handling of effective teams requires a focus on functional leadership in team management. A review of team dynamics, from both the internal and external perspectives, should provide data that can be captured for the edification of future team leaders and the improvement of project productivity. This was accomplished through the development and administration of questionnaire surveys and personal interviews to validate perceptions. An emphasis on the comparative approach was taken to evaluate production results of teams and traits of effective team leaders to those less effective. Important variables that skew the findings within this unique work environment were taken in consideration.

Lastly, it was the aim of this study to provide a better understanding by both clients and workers of the various factors that transform a seemingly simple UXO clearance project into a complex endeavor. Every bomb dropped that does not function as designed throughout the various world conflicts allows for expansion of the UXO industry.

Definition of Terms

The following terms are used throughout this study:

1. Demilitarization. The mutilation or disfigurements of ammunition, equipment, or material to prevent further use.

2. Detonation. A violent chemical reaction within a chemical compound or mechanical mixture involving heat and pressure.

3. EOD. Explosive Ordnance Disposal.

4. Explosive. A substance or mixture of substances, which under external influences, rapidly releases energy in the form of gases and heat as to be capable of causing damage to the surroundings.

5. Group. Two or more persons interacting in such a manner that each person influences and is influenced by each other person, and who may or may not have unanimity of purpose (Dessler, 2001, p. 551).

6. Leadership. The process of influencing others to facilitate the attainment of organizationally relevant goals (Ivancevich & Matteson, 2002, p. 425).

7. Management. The process of planning, organizing, leading, and controlling the use of resources to accomplish performance goals (Schermerhorn, 2002, p. 20).

8. Manager. A person who plans, organizes, leads, and controls the work of others so that the organization achieves its goals (Dessler, 2001, p. 3).

9. Non-UXO related materials. Man made materials, i.e. plastic, fence posts, wood posts, etc.

10. OE. Other explosive.

11. Ordnance. Military material including ammunition and equipment required for their use.

12. Target materials. Items used as military targets.

13. Team. A group of people committed to a common purpose, set of performance goals, and approach for which they hold themselves mutually accountable (Dessler, 2001, p. 557).

14. Thermal processing. The process of exposing materials to a high temperature to decontaminate and release them without restrictions.

15. TPU. Thermal Processing Unit.

16. UXO. Unexploded ordnance.

17. UXO escort. An individual that is a graduate of the Naval School Explosive Ordnance Disposal.

18. UXO-related remnants. Remnants that originated from UXO including all exploded remnants recovered during clearance operations.

19. UXO specialist. An individual that is a graduate of the Naval School Explosive Ordnance Disposal and has a minimum of five years of experience in the UXO field.

20. UXO supervisor. An individual that is a graduate of the Naval School Explosive Ordnance Disposal and has a minimum of seven years of experience in the UXO field.

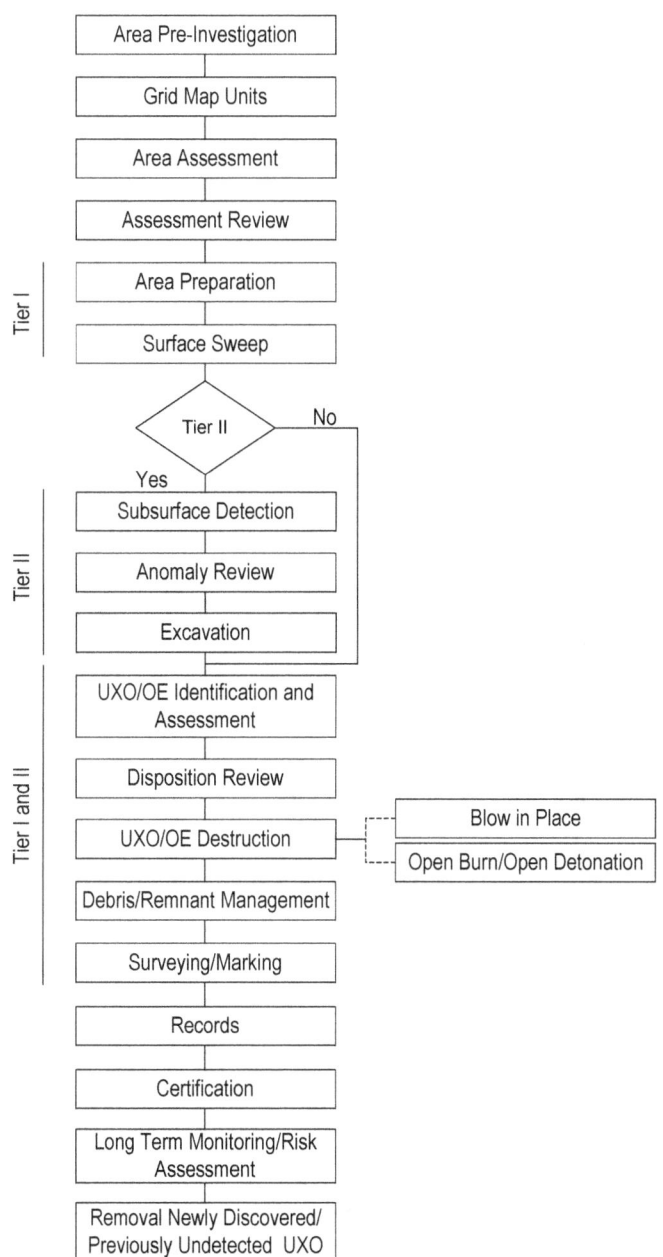

Figure 1 UXO Clearance Process

Chapter 2

A REVIEW OF RELATED LITERATURE

In the extensive literature that is available on the topic of leadership, a number of approaches with many variations and different areas of emphasis have been established. Many models have been developed to capture the essence of the relationship between leadership and effectiveness. While no one model has dominated the field of study, it is important to review and understand the most common of the models as a foundation for the understanding of the important relationship between leadership and effectiveness.

This review of literature is intended to provide highlights in the existing body of literature that are most relevant to this research. It begins with a discussion of teams versus groups and what leadership is. Various approaches, theories and models of leadership follow the introductory discussion. Then, the focus of the research becomes more selective and focuses on the specific application of leadership and skills as they apply to the team organizational context. The final chapter concludes with an examination of how leadership effectiveness is assessed within a team.

Teams versus Groups

Oftentimes the terms teams and groups are used interchangeably in the literature. However, they are not the same (Ivancevich & Matteson, 2002, p. 312). Dessler (2001) shares the differences between these two terms. Guzzo and Dickson (1996) assert that a team is more complex than a group. In order to gain the benefits derived from teamwork, the differences between teams and groups need to be understood. Table I below summarizes the differences.

Formal Work Group	Team
Works on common goals	Total commitment to common goals
Accountable to manager	Accountable to team members
Skill levels are often random	Skill levels are often complementary
Performance is evaluated by leader	Performance is evaluated by members as well as leaders
Culture is one of change and conflict	Culture is based on collaboration and total commitment to common goals
Performance can be positive, neutral, or negative	Performance can be greater than the sum of members contributions or synergistic (e.g. 1 + 1 + 1 = 5)
Success is defined by the leader s aspirations	Success is defined by the members aspirations

Table 2 Comparison of Groups and Teams
SOURCE: J. M. Ivancevich & M. T. Matteson, *Organizational behavior and management,* p. 313. Copyright © 2002 by McGraw-Hill. Reprinted by permission of The McGraw-Hill Companies.

The term team will be used to refer to the work groups that were surveyed in this study. It is also understood that leadership occurs in groups and that this approach is consistent with team leadership theory and research (Northouse, 2001).

What is Leadership?

Even though there is a tremendous interest in leadership today, agreement on the definition of leadership is varied and has resulted in the investigation of many different aspects of leadership (Hughes, Ginnett, & Curphy, 2002; Ivancevich & Matteson, 2002; Lord, 1977; Northouse, 2001; Schermerhorn, 2002; Schermerhorn, Hunt, & Osborne 2000). Bass and Stogdill (1990) state: There are as many different definitions of leadership as there are persons who have attempted to define the concept (p. 11). Many researchers define leadership from the perspective of power; others on the ability to set direction and to influence others to follow the desired direction. It is important to recognize that there is no single definition that is more accurate than the next (Hughes, 2002). Due to the varied organizational contexts and situations that require leaders to demonstrate leadership, Ivancevich s et al. broad definition has been adopted in this research as being the most comprehensive and as the process of influencing others to facilitate the attainment of organizationally relevant goals (Ivancevich et al., 2002, p. 425).

Despite the differences in the definition of leadership, agreement on several aspects of leadership is shared by many of the researchers. Northouse (2001) views these common elements as: (a) leadership is a process, (b) leadership involves influence, (c) leadership occurs within a group context, and (d) leadership involves goal attainment (p. 3). Bennis and Biederman (1997) further refine the common aspects for effective leaders by concluding that:

1. They provide direction and meaning to the people that they are leading. This means they remind people what is important and what they are doing makes an important difference.
2. They generate trust.
3. They favor action and risk taking. That is, they are proactive and willing to risk failing in order to succeed.
4. They are purveyors of hope. In both tangible and symbolic ways they reinforce the notion that success will be attained.

Csoka (1998) notes that today CEO growth challenges have less to do with managing organizational assets and more to do with developing quality leadership, workforce performance, and relationships at all levels of an organization (p. 8).

Leadership and Management

There exists controversy in the literature as to whether leaders and managers and leadership and management are different (Bass et al., 1990; Hughes et al., 2002; Northouse, 2001; Schermerhorn et al., 2000).

The traditional view of management associates it with the four basic functions of planning, organizing, leading, and controlling (Dessler, 2001). Schermerhorn (2002) defines management as the process of planning, organizing, leading, and controlling the use of resources to accomplish performance goals (p. 20).

Bass et al. (1990) reflect that leaders manage and managers lead, but the two activities are not synonymous (p. 383). Bennis and Nanus (1985) make the distinction between managers and leaders by viewing mangers as people who do things right and leaders as people who do the right thing (p. 21). Bryman (1986) states the differences based on the relation of time with leadership reflecting the desired future state and management with the preoccupation with the here-and-now of goal attainment (p. 6). Other viewpoints stipulate that a manager may not necessarily be a leader and vice versa (Kotter, 1992; Yukl, 1989). Northouse (2001) presents the distinction that management produces order and consistency, whereas leadership produces change and movement. Hughes et al. (2002, p.10) and Northouse (2001, p. 9) address a comparison of the distinctions between management and leadership that have been consolidated and reflected in Table 3 below.

Yukl (1989) summarizes the distinction in the literature by concluding, Leaders influence commitment, whereas managers merely carry out position responsibilities and exercise authority. Additionally,

Yukl (1989) recognizes that the arguments revolve around the degree of

overlap between managing and leading. For the purposes of this

dissertation, these distinctions will not be emphasized and the terms will

be used interchangeably. After all, the focus of this research is on the

process of achieving goals by influencing a team. Both management and

leadership require the influencing of a group towards a goal (Northouse,

2001). The most direct influence of leadership happens at the team

supervisor level where the success or failure of a team is determined on

the effectiveness of the team leader s leadership abilities (Csoka, 1998).

Table 3 Comparisons of Management and Leadership

MANAGEMENT	LEADERSHIP
Managers administer	Leaders innovate
Managers control and problem solve	Leaders motivate and inspire
Managers maintain	Leaders develop
Managers have a short-term view	Leaders have a long-term view
Managers ask how and when	Leaders ask what and why
Managers imitate	Leaders originate
Managers accept the status-quo	Leaders challenge the status quo
Managers plan and budget	Leaders build a vision and strategize
Managers organize and staff	Leaders align people and communicate

Trait Approach to Leadership

Numerous theories have been established to better understand leadership. Much of the early research focused on determining whether leaders were born or made. The pursuit of the superior qualities and individual traits that differentiated a leader from his followers was the objective of the trait theories (Bass et al., 1990). The Great Man Theory reflected this difference between the leader and the follower (Schermerhorn, et al. 2000). However, this theory was later discounted due to lack of support (Dimock, 1970a). Bass et al. (1990) conclude that personality traits differentiate leaders from followers, successful from unsuccessful leaders, and high-level from low-level leaders. The trait approach to personality maintains people behave the way they do because of the strengths of the traits that they possess (Hughes et al., 2002, p. 168). Additionally, the situation does have an influence on who is successful and effective (Bass et al, 1990; Dimock, 1970a).

Table 4 below summarizes the traits most commonly found in successful leaders. It is important to note that leadership effectiveness cannot be determined absolutely from these or other traits (Dimock, 1970a; Ivancevich et al., 2002). Bryman (1986) indicates that this inability to be able to differentiate between effective and ineffective leaders based on traits resulted in the new focus of behavior analysis.

Intelligence	Personality	Abilities
Judgment	Adaptability	Ability to enlist cooperation
Decisiveness	Alertness	Cooperativeness
Knowledge	Creativity	Popularity and prestige
Fluency of speech	Personal integrity Self-confidence Emotional balance and control Independence (nonconformity)	Sociability (interpersonal skills) Social participation Tact, diplomacy

Table 4 Traits Associated with Leadership Effectiveness
SOURCE: J. M. Ivancevich & M. T. Matteson, *Organizational behavior and management,* p. 429. Copyright © 2002 by McGraw-Hill. Reprinted by permission of The McGraw-Hill Companies.

Dessler (2001) identifies six traits that differentiate an effective leader from an ineffective leader: drive, the desire to lead, honesty and integrity, self-confidence, cognitive ability, and knowledge of the business (pp. 296-297). It is interesting to note that knowledge is a common trait to both Ivancevich et al. and Dessler s list; however, exceptions do occur with a CEO not necessarily possessing the specific knowledge unique to a company but having the business acumen to function effectively.

Two of the weaknesses of the trait approach include lack of a comprehensive list of consistent patterns of leadership traits that has yet to be determined and the degree of subjectivity involved in determining the most important leadership trait (Northouse, 2001). Bass et al. (1990) add, The trait approach ignored the interaction between the leader and his or her group (p. 511). The trait approach is viewed as not being effective as a leadership predictor (Ivancevich et al., 2002); however, it can be beneficial in unfamiliar, ambiguous, or what we might call weak situations (Hughes et al., 2002).

The Five Factor Model (FFM) of personality categorizes common traits into five broad personality dimensions as depicted in the Table 5 below. Researchers have validated leadership success as being positively correlated with the FFM personality dimensions (Hughes et al., 2002).

Five Factor Dimension	Traits	Behaviors/Items
Surgency	Dominance Sociability	I like having responsibility for others. I have a large group of friends.
Agreeableness	Empathy Friendly	I am a sympathetic person. I am usually in a good mood.
Dependability	Organization Credibility Conformity	I usually make to do lists. I practice what I preach. I rarely get into trouble.
Adjustment	Steadiness Self-acceptance	I remain calm in pressure situations. I take personal criticism well.
Intellectance		I like traveling to foreign countries.

Table 5 The Five Factor Model of Personality
SOURCE: R. L. Hughes, R. C. Ginnett, & G. J. Curphy, *Leadership: Enhancing the lessons of experience,* p. 171. Copyright © 2002 by McGraw-Hill. Reprinted by permission of The McGraw-Hill Companies.

The most popular personality test is the Meyers-Briggs Type Indicator (MBTI), which evaluates a person s preferences on four dimensions to determine a psychological type. Leaders were categorized predominantly as thinking and judging over feeling and perceiving

(Hughes et al., 2002). This test needs to be used as a guide rather than an absolute because types are not completely stable over time and type categorizing oftentimes results in classification profiling of personnel (Hughes et al., 2002).

An important individual variable of a leader s personal skills with implications for leadership is emotional intelligence (EI). While there is debate as to the exact definition, Ivancevich et al. (2002) define emotional intelligence as the ability of people to understand and manage their personal feelings and emotions, as well as their emotions toward other individuals, events, and objects (p. 673). Researchers agree that emotional intelligence involves abilities in five areas: self-awareness, self-regulation, motivation, empathy, and social skill (Dessler, 2001; Hughes et al., 2002; Ivancevich et al., 2002; Schermerhorn, 2002). Emotional intelligence is important to leadership because as Hughes et al. report: Leaders who can better align their thoughts and feelings with their actions may be more effective than leaders who think and feel one way about something but then do something different about it (p. 202).

Style Approach to Leadership

As researchers shifted their focus from the trait approach and a leader s attributes to a leader s actions, the style approach, which is also commonly referred to as the behavior approach to leadership, emerged. It is usually viewed from two perspectives: accomplishing the task or task behaviors and satisfying the needs of team members or relationship behaviors (Dessler, 2001; Northouse, 2001). The evaluation of behavior through observation is easier than defining personality traits, which requires the inference of behavior or measurement with tests (Hughes et al., 2002).

Leadership research during the Ohio State Studies developed the first form of the Leadership Behavior Description Questionnaire by initially collecting 1,800 questionnaire items and consolidating them to 150 items to rate a leader s performed behaviors as seen by his or her subordinates (Bass et al., 1990; Hughes et al., 2002; Northouse, 2001). The results of this study demonstrated that the subordinates responses identified two primary leadership behaviors that were distinct and unrelated: consideration and initiation of structure (Bass et al., 1990; Dessler, 2001; Hughes et al., 2002; Northouse, 2001). Consideration was a relationship behavior by a leader towards his or her subordinates and reflects the level of concern for the welfare of the other members (Bass et al., 1990; Hughes et al., 2002; Northouse, 2001). These studies also determined that leaders who were above average in leadership show more empathy or understanding of others and are also more accurate in ability to judge other members feelings and opinions on issues relevant to

the group (Dimock, 1970a, p. 9). Initiation of structure was a task

behavior by the leader and shows his or her emphasis on initiating work,

organizing work, giving structure to the work context, defining role and

responsibilities, and scheduling work activities (Bass et al., 1990; Hughes

et al., 2002; Northouse, 2001). However, Bryman (1986) views one of the

shortcomings of this study as the inability to differentiate between the

influences of leadership versus management. The discovery of the

optimum mix of these two behaviors, consideration and initiation of

structure, to determine effectiveness continues to be a task of researchers

today (Northouse, 2001).

Researchers at the University of Michigan identified two leadership

behaviors exhibited in work settings: employee orientation and

 production orientation (Dessler, 2001; Hughes et al., 2002; Northouse,

2001). Employee orientation is very similar to the consideration behavior

and production orientation to the initiation of structure behavior found in

the Ohio State Studies (Hughes et al., 2002; Northouse, 2001). The

fundamental difference between the Ohio State and the University of

Michigan Studies centered around the assumption of these behaviors with

Ohio State viewing them as independent whereas the University of

Michigan as mutually exclusive (Hughes et al., 2002). However, the key assumption common to both focused on a leader s ability that behavior could influence a group to goal accomplishment (Hughes et al., 2002).

The results of behavior research suggest that effective leaders are concerned with both people and task or production (Schermerhorn, 2002). The Blake and Mouton s Leadership Grid profiles leader behavior on these two dimensions (Hughes et al., 2002). However, additional research studies reveal that the context and style of a leader s behavior and the criteria used to judge effectiveness also influence the assessment of leadership effectiveness (Hughes et al., 2002).

Several weaknesses are associated with the style approach to leadership. The first criticism is that the style approach has not established one style that is common to all situations (Northouse, 2001). Yukl (1988) indicates that behavior varies from context to context and leadership effectiveness varies across situations.

Situational Leadership Theory

The situational leadership theory focuses on the context of the situation. However, situations are oftentimes complex and most of the time unique, so the influence on leadership is variable. Consequently, an effective leader needs to adapt his or her style to the demands of different situations (Northouse, 2001). Additionally, leadership is contingent upon the dynamics of the leader, the follower, and the situation (Hughes et al.,

2002). Therefore, the importance of a leader s diagnostic ability cannot be

overemphasized (Hersey, Blanchard, & Johnson, 2001, p. 171).

As this theory evolved, the previous focus on the leader behaviors

of initiating structure and consideration from the Ohio State studies were

modified with a focus on the amount of direction (task behavior) and the

amount of socio-emotional support (relationship behavior) a leader must

provide given the situation (Hughes et al., 2002). Hersey et al. (2001)

define task behavior and relationship behavior as follows (p. 173):

> *Task behavior* is defined as the extent to which the leader engages
> in spelling out the duties and responsibilities of an individual or the
> group. These behaviors include telling people what to do, how to do
> it, where to do it, and who is to do it.

> *Relationship behavior* is defined as the extent to which the leader
> engages in two-way or multiway communication. The behaviors
> include listening, facilitating, and supportive behaviors.

Additionally, the Ohio State studies revealed that leadership styles varied

and that these two dimensions were not consistently related to leadership

success; the relative effectiveness of these two behavioral dimensions

often depended on the situation (Hughes et al., 2002). Furthermore, it

was determined that these two behavioral dimensions are separate and

distinct dimensions (Hersey et al., 2001).

Hersey et al. (2001) found the following situational factors as influencing leader effectiveness (p. 175):

> Leader
> Followers
> Supervisor
> Key associates
> Organization
> Job demands
> Decision time

To continue, Hersey et al. (2001) reflect, the relationship between leaders and followers is the crucial variable in the leadership situation (p. 175). It is this collaboration of the leader and the follower that creates a synergy that results in achievements that could not have been accomplished as individuals and provides satisfaction to both (Bennis & Biederman,1997).

In a refinement to the original Situational Leadership Theory, Hersey and Blanchard added a curved line that represents the leadership behavior that will most likely be effective given a particular level of follower maturity (Hughes et al., 2002). Maturity is viewed from two components: job maturity or self-confidence, ability, and readiness to accept responsibility and psychological maturity or relevant skills and technical knowledge (Hughes et al., 2002). Northouse (2001) reflects that (p. 58):

> Employees are at a high development level [maturity] if they are interested and confident in their work and they know how to do the task. Employees are at a low development [maturity] if they have little skill for the task at hand but feel as if they have the motivation or confidence to get the job done.

It is important to note that both of these components of maturity are meaningful only with regard to a particular task (Hughes et al., 2002).

The Situational Leadership Theory reflects strong direction (task behavior) is appropriate with followers with low readiness   and an increase in readiness on the part of people who are somewhat unready should be rewarded by increased positive reinforcement and socioemotional support (relationship behavior) (Hersey et al., 2001, p. 190). Conversely, as the subordinate maturity increases, leadership should be more relationship-motivated than task motivated. This model addresses leader behaviors to four degrees of subordinate maturity levels from highly mature to highly immature; leadership can consist of the following behaviors towards the subordinate: delegating,   participating, selling, and telling (Hersey et al., 2001; Schermerhorn, 2002).

Northouse (2001) identifies the following strengths in the Situational Leadership Theory: stood the test in the marketplace, practicality, prescriptive value, emphasizes the concept of situational leadership, reminds us to treat each subordinate differently based on the task at hand (pp. 60-61). The weaknesses pointed out by Hughes et al. (2002) are summarized below (pp. 367-368):

Little research to support the predictions

Inadequate rationale or sufficiently specific guidance about why or how particular levels of task and relationship behaviors correspond to each of the follower maturity levels No evidence that leaders who behave according to the models prescriptions actually have higher unit performance indexes, better performing or more satisfied subordinates, or a more favorable organizational climate

Northouse (2001) adds one more criticism by stating that the model does not provide guidelines on application of the model in a group versus a one-to-one context.

Leadership Contingency Model

Oftentimes the Leadership Contingency Model is viewed to be almost the opposite of the Situational Leadership Theory. Table 6 depicts these differences.

Situational Leadership Theory (SLT)	Leadership Contingency Model
Emphasis is on flexibility in leader behaviors.	Leaders are much more consistent and less flexible in their behaviors.
Leaders who correctly base their behaviors on follower maturity will be more effective.	Leader effectiveness is primarily determined by selecting the right kind of leader for a certain situation or changing the situation to fit the particular leader s style.

Table 6 Comparison of SLT to Contingency Theory
SOURCE: R. L. Hughes, R. C. Ginnett, & G. J. Curphy, *Leadership: Enhancing the lessons of experience,* p. 369. Copyright © 2002 by McGraw-Hill. Reprinted by permission of The McGraw-Hill Companies.

Fiedler s Leadership Contingency Model suggests that three major situational variables determine whether a given situation is favorable to leaders with high levels of these three factors giving the most favorable situation and low levels the least favorable. (Hersey et al., 2001, p. 110):

1. Their personal relations with the members of their group (leader-member relations)
2. The degree of structure in the task that their group has been assigned to perform (task structure)
3. The power and authority that their position provides (position authority)

Oftentimes the effectiveness of leaders will vary between each other in similar situations (Hughes et al., 2002). The leader-member relations is the most influential variable in determining favorability of a situation (Hughes et al., 2002).

Fiedler developed the Least-Preferred-Coworker (LPC) scale to measure leadership effectiveness (Hughes et al., 2002; Northouse, 2001). Both Hughes et al. (2002) and Northouse (2001) reflect that leaders who score high on the LPC scale are described as relationship motivated, and those who score low as task motivated. With further analysis, Fiedler concluded the following (Hersey et al., 2001, p. 110):

1. Task-oriented leaders tend to perform best in group situations that are either very favorable or very unfavorable to the leader.
2. Relationship-oriented leaders tend to perform best in situations that are intermediate in favorableness.

Fiedler suggests that it may be easier for leaders to change their situation to achieve effectiveness, rather than change their leadership style (Ivancevich et al., 2002).

The Contingency Leadership Theory is supported by extensive research (Hersey et al., 2001; Hughes et al., 2002; Northouse, 2001). Additionally, the theory has highlighted the importance of situational considerations on leaders (Hersey et al., 2001; Hughes et al., 2002; Northouse, 2001). A review of the literature reveals the following summarized review of the shortcomings of the theory (Hughes et al., 2002; Ivancevich et al., 2002; Northouse, 2001):

1. Field settings by other researchers concluded in mixed results.

2. There were too many uncertainties regarding the measurement and interpretation of LPC scores and the scale.

3. The interpretation of situational favorability varies according to the effectiveness of leadership styles.

4. The relationships between LPC scores and situational favorability are unclear.

5. The theory does not address how to remedy mismatches between a leader and a situation.

Path-Goal Theory

The Path-Goal Theory is based on similar concepts addressed by the Ohio State leadership studies with the two dimensions of initiating structure and consideration and the Expectancy Model of Motivation (Hersey et al., 2001; Hughes et al., 2002; Northouse, 2001). The Expectancy Model of Motivation focuses on effort-performance and the performance-goal satisfaction (reward) linkages (Hersey et al., 2001; Hughes et al., 2002). The effective leader provides guidance, motivation, and support through task and relationship leadership to enhance the attainment of goals (Hughes et al., 2002; Northouse, 2001). Hersey et al. (2001) point out that leaders are most successful when they supply what is missing from the situation (p. 112). The Path-Goal Theory suggests that the following four leadership behaviors be used in different situations to enhance satisfaction, effective performance and rewards: directive leadership, supportive leadership, participative leadership, and achievement-oriented leadership (Hughes et al., 2002; Northouse, 2001).

Northouse (2001) identifies three strengths of Path-Goal Theory (pp. 96-97):

1. It provides a useful theoretical framework for understanding how various leadership behaviors affect the satisfaction of subordinates and their work performance.
2. It attempts to integrate the motivation principles of expectancy theory into a theory of leadership.

3. It provides a model that in certain ways is very practical by reminding leaders that the overarching purpose of leadership is to guide and coach subordinates as they move along the path to achieve a goal.

Weaknesses of the Path-Goal Theory are summarized as follows (Hughes et al., 2002; Northouse, 2001):

1. Mixed support by researchers

2. Very complex and can be confusing to interpret

3. Does not adequately address relationship between leadership behavior and motivation

4. Does not consider selection and training of followers

Leader-Member Exchange Theory

The Leader-Member Exchange (LMX) Theory focuses on the exchange between leaders and followers (Hughes et al., 2002; Northouse, 2001). This theory suggests that leadership develops over time in three phases: the stranger phase, the acquaintance phase, and the mature partnership phase (Northouse, 2001). Additionally, it suggests that leaders group subordinates into either in-group or out-group members (Ivancevich et al., 2002).

The Leader-Member Exchange Theory reflects the importance of communications and indicates Effective leadership occurs when the communication of leaders and subordinates is characterized by mutual trust, respect, and commitment (Northouse, 2001, p. 119). Two shortcomings of this theory are that it appears to favor discrimination and it is not fully supported (Northouse, 2001).

Vroom-Yetten Contingency Model

This model is also sometimes referred to as the Vroom-Jago Leadership Model. Vroom-Yetten developed a normative model that is designed to provide guidelines to a leader to assess a problem, take in account the various situational variables, and choose the most appropriate decision-making style (Hersey et al., 2001; Ivancevich et al., 2002; Schermerhorn, 2002). Their approach assumed that flexibility to situations was necessary and that an effective leader is that person who consistently chooses the most appropriate decision for a situation (Ivancevich et al., 2002; Schermerhorn, 2002). Gist, Locke, & Taylor (1987) indicate the decision to allow participation should be based on such considerations as the need for a high quality decision, the boss's knowledge, and the probability that the decision will be accepted if it is made by a superior (p. 246).

Five leadership styles are identified for making a decision within an individual or group situation (Hersey et al., 2001; Ivancevich et al., 2002):

Autocratic I: The leader makes the decision to solve the problem based on information already available.

Autocratic II: The leader obtains additional information from the group before making a decision.

Consultative I: The leader shares the problem with subordinates individually before making a decision.

Consultative II: The leader shares the problem with subordinates as a group before making a decision.

Group II: The leader shares the problem with subordinates as a group and the group makes the decision with the leader acting as the chair.

A situational diagnosis consisting of seven items is suggested before the selection of the most appropriate decision-making style (Hersey et al., 2001; Ivancevich et al., 2002). The first three items pertain to the quality or technical accuracy of the decision; the last four pertain to the acceptance of the decision style (Hersey et al., 2001; Ivancevich et al., 2002).

Hersey et al. (2001) identify three reasons to view this model as an important theory: it is a widely respected leadership behavior theory that has major support among the contingency theories, it provides the leader with the flexibility to adapt the best approach to a situation, and it affords the opportunity for leaders to develop into more effective leaders.

However, Hughes et al. (2002) and Ivancevich et al. (2002) criticize this theory as lacking validity due to insufficient evidence in that it does not unequivocally prove that those that follow the model are more effective than those that do not and it oversimplifies decision-making by viewing it a single point in time.

Transformational Leadership

Transactional leadership focuses on this exchange between the leader and subordinate for achieving routine performance and is the subject of the majority of leadership models (Bass et al., 1990; Dessler, 2001; Hughes et al., 2002; Ivancevich et al., 2002; Northouse, 2001; Schermerhorn, 2002; Schermerhorn et al., 2000). Bryman (1986) views this interface between the leader and the subordinate as a two-way influence process that ultimately results in the leader influencing the subordinate somehow (p. 7). Many researchers viewed the interpretation of leadership from the transactional perspective as limiting and not encompassing the full concept of leadership (Bass et al., 1990). Bass et al. state that:

> The transformational leader asks followers to transcend their own self-interests for the good of the group, organization, or society; to consider their longer-term needs to develop themselves, rather than their needs of the moment; and to become more aware of what is really important. (Bass et al., 1990, p. 53)

Four dimensions are associated with transformational leadership (Dessler,
2001; Hughes et al., 2002; Ivancevich et al., 2002; Northouse, 2001;
Schermerhorn, 2002; Schermerhorn et al., 2000):

Charisma   leaders who exhibit high standards, provide a vision
and mission to followers, and arouse follower respect and trust

Inspiration   leaders who inspire and motivate followers through
communication to achieve high standards towards the attainment of
the vision

Intellectual stimulation   leaders who stimulate followers to become
creative and innovative towards problem solving

Individualized consideration   leaders who treat followers as
individuals and advise and coach them towards success

It is important to note that both transformational and transactional
leadership are used concurrently and are not mutually exclusive
(Schermerhorn et al., 2000). In fact in those situations in which change is
necessary, the combination is necessary to achieve sustainable high-
performance results (Schermerhorn, 2002, p. 351). Two additional
dimensions unique to transactional leadership added during those
situations in which a combination of both leadership approaches are used
are (Ivancevich et al., 2002; Northouse, 2001):

Contingent reward   leaders who reward followers for goal

accomplishment as mutually agreed upon

Management by exception   leaders who allow followers to perform

and intervene only when necessary to provide corrective action or

non-achievement of standards

Transformational leadership is widely researched from many

different perspectives, typifies the leadership exhibited by the ideal

leader, expands the leadership concept presented by the transactional

approach, and focuses on the needs of the follower (Bass et al., 1990;

Hughes et al., 2002; Northouse, 2001). However, Northouse (2001)

expresses the following criticisms of the transformational leadership

approach which include:

It is difficult to interpret clearly because it is a very broad approach

that is difficult to succinctly define and oftentimes is viewed from a

simplistic perspective.

Because of its focus on leader dimensions, it emphasizes a trait

focus instead of also considering the behavioral element.

It establishes a potential of abusing subordinates through control

and power.

Ginnett s Team Effectiveness Leadership Model

This model was developed to assist the team leader in identifying

problem areas or improving team effectiveness. It takes a systems

theory approach by incorporating the inputs of the team, processing the

inputs, and generating output as a function of the team s effectiveness

(Hughes et al., 2002). Inputs include various factors, for instance

individual, organizational, and team design, that all have an influence on

team performance (Hughes et al., 2002). The process phase reflects how

the team performs work, which can be measured by determining whether

a team:

> Works hard enough
> Has sufficient knowledge and skills within the team to perform the
> task
> Has an appropriate strategy to accomplish its work
> Has constructive and positive group dynamics among its members
> (Hughes et al., 2002, p. 309)

The output is the actual team production. Hughes et al. (2002) reflect an

argument that a team is effective if:

> The team s productive output (good, services, decisions) meets the
> standards of quantity, quality, and timeliness of the people who use
> it
> The group process that occurs while the group is performing its
> task enhances the ability of the members to work as members of a
> team (either the one they were on or any new teams they may be
> assigned to) in the future.
> The group experience enhances the growth and personal well-
> being of the individuals who compose the team.
> (Hughes et al., 2002, p. 307)

This model is not well covered by the literature. Hughes et al. (2002) note that this model is too mechanistic and simplistic in its consideration of the various variables that could affect a team s effectiveness, but yet has been adopted by many educators as a tool to help understand and teach team effectiveness.

Team Leadership Theory

The Team Leadership Theory concentrates on what makes teams effective or what constitutes team excellence (Northouse, 2001, p. 166). Hackman and Walton (1986) identify five components necessary for effectiveness; Larson and LaFasto (1989) identify eight. There appears to be consistency between the two researchers. A comparison of theories is depicted in Table 7 below.

Goals need to be clear and motivating so that a determination can be made as to the fulfillment of the goal, and a sense of success is achieved (Larson et al., 1989; Northouse, 2001). Clarity implies that there is a specific performance objective (Larson et al., 1989, p. 28). Team failure oftentimes is due to a distraction from the team goal and a result of conflicting personal objectives and political motivations (Larson et al., 1989).

Conditions of Group Effectiveness (Hackman & Walton, 1986)	Characteristics of Team Excellence (Larson & LaFasto, 1989)
Clear, engaging direction	Clear, elevated goal
Enabling structure	Results-driven structure Competent team members Unified commitment Collaborative climate
Enabling context	Standards of excellence
Expert coaching	Principled leadership
Adequate material resources	External support

Table 7 Comparison of Theory and Research Criteria
SOURCE: P. G. Northouse, *Leadership: Theory and practice*, p. 167. Copyright © 2001 by Sage Publications, Inc. Reprinted by permission of Sage Publications, Inc.

Team structure is important to goal achievement (Larson et al., 1989; Northouse, 2001). The focus of this paper concentrates on UXO work teams that are production oriented and rely on technology and labor. Larson et al. (1989) classify the objective of the UXO team as tactical with a process emphasis on directive, highly focused tasks, role clarity, well-defined operational standards, and accuracy (p. 43). Additionally, all teams, in order to be effective, need to have well-defined position descriptions with a clear understanding of the responsibilities associated with each position, good communications, methods to assess individual performance and provide evaluation, and focus on objective decision making (Larson et al., 1989).

A team needs to consist of individuals that possess the right technical and personal skills to be competent in their assigned job (Larson et al., 1989; Northouse, 2001). Hackman et al. (1986) argue that a team member needs to be provided adequate training and education to function effectively. Larson et al. (1989) present the dominant selection criteria of a tactical team member to be loyal, committed, action-oriented, responsive, and pressing (p. 67). Three common traits of competent team members of effective teams include essential skills and abilities, strong desire to contribute, and effective collaboration (Larson et al., 1989, pp. 69-70).

A common sense of unity or identification needs to be developed within a team (Northouse, 2001, p. 169). Lack of a unified commitment is often the most clearly missing feature of ineffective teams (Larson et al., 1989, p. 73). Guzzo et al. (1996) agree with the finding that conflict between individual and team goals result in team dysfunction. Team effort, including time and energy, is the minimum essential ingredient to success with teamwork being the result of team identification and unity (Larson et al., 1989).

Trust through honesty, openness, consistency, and respect establishes a climate that fosters collaboration between team members

(Larson et al., 1989, p. 85). Larson and LaFasto (1989) note that

teamwork results from collaboration because trust:

> Allows team members to stay problem-focused
> Promotes more efficient communication and coordination
> Improves the quality of collaborative outcomes
> Leads to compensating
> (Larson et al., 1989, pp. 88-92)

Hackman et al. (1986) indicate that performance needs to be

evaluated so that progress can be noted and tasks accomplished.

Northouse (2001) reveals that a team needs to identify and establish

standards of performance so that pressures can drive members to perform

at the highest level; otherwise, performance excellence will not be

achieved. Pressures to achieve standards of excellence can occur

different ways and from various sources:

> Individual standards consist of those performance expectations that
> each member of the team embraces as personal pressure to
> achieve.
> Team pressure, similar to that exerted by individuals, can
> eventually influence individual performance.
> The consequences of success or failure can exert pressure to
> create standards that make success most predictable.
> External pressure consists of any impelling or constraining forces
> outside the actual team that exert influence on the team s
> performance.
> The team leader represents the final source of pressure to perform.
> (Larson et al., 1989, pp. 96-99).

Teams are frequently tasked with challenging assignments, but are not

supported nor rewarded for their efforts (Northouse, 2001). The lack of external support and recognition appears to be more of a factor for ineffective teams than an asset for effective teams (Larson et al., 1989). Crosby (1986) adds that it is most important to ensure that the leader provides recognition to those that have earned it. Excellence can be achieved by associating rewards collectively to team success instead of individual achievement (Larson et al., 1989).

Effective leaders are instrumental to the success of a team. Larson and LaFasto (1989) have found three characteristics that are exhibited by effective leaders: establish a vision in a manner that fosters shared commitment towards the goal, create change towards constant improvement, and unleash talent through inspiration and motivation (p. 121). Leaders that establish a collaborative climate that focuses on shared instead of individual contributions are helpful to goal achievement and can bring synergy to knowledge levels (Northouse, 2001). Katzenbach and Smith (1992) describe the actions required of the team leader to orchestrate these efforts by encouraging individual creativity, but yet focusing on the collective contribution of goal accomplishment as maintaining a "delicate balance" of team leadership (p. 128). The establishment of shared principles identifying clear expectations of team members and the team leader can serve as a guide in the establishment of standards of performance, maintain a reasonable goal focus, and create an appropriate decision-making climate (Larson et al., 1989). The following five principles of team leader behavior were found to be key in unleashing team member talent and establishing a positive and supportive decision-making climate:

> Trusting team members with meaningful levels of responsibility.
> Providing team members the necessary autonomy to achieve results.
> Presenting challenging opportunities, which stretch the individual abilities of team members.
> Recognizing and rewarding superior performance.
> Standing behind our team and supporting it.
> (Larson et al., 1989, p. 126)

Effective leaders inspire team members to perform at the highest levels and exhibit their own leadership (Larson et al., 1989). Often it has been said that Leaders beget leaders! Kelley (1998) theorizes that, Boosting the productivity of the entire brain-powered workforce, not just depending on one or two stars to carry the team, is the key to success (p. 12). Antonioni (1994) adds that leaders need to become adept in five different styles of management as they lead teams to success: (1) internal consultant, (2) visionary, (3) experimenter, (4) coach, and (5) educator. Crosby (1986) concludes that a leader must enhance team member conviction of desired goal accomplishment, serious commitment to the team and goal, and conversion in the soul of the team to increase productivity (pp. 78-80). Laroche (2001) presents the multi-cultural dimension that could affect internal team dynamics and the approach that a leader uses based on different acceptable cultural values.

Team Leadership Model

The Team Leadership Model integrates leader mediation decisions with internal and external leadership functions and focuses on the outcome of team effectiveness from the perspectives of performance and team development (Northouse, 2001). The model is based on the functional actions required by the leader through corrective actions or suggestions to ensure team effectiveness (Northouse, 2001). Monitoring behaviors are used to assess and improve performance and action behaviors to consider the influences of the internal and external factors (Gist et al., 1987).

Northouse (2001) identifies the following strengths of this model: a diagnostic approach with a focus on performance and team effectiveness, a team leader s guide to team design and effectiveness, and objective criteria for team leader selection. Weaknesses include the complexity of the model, vague leader decision options, and the need for further validation (Gist et al., 1987; Northouse, 2001).

Leadership Effectiveness

The approach to achieving leadership effectiveness varies almost as much as the definition of "leadership." Team members perceive internal factors to be influential and contribute significantly towards team effectiveness (Kline & MacLeod, 1996). Fiedler (1958) takes the strict interpretation and relates leadership effectiveness to successful task accomplishment, but recognizes that sometimes a team achieves effectiveness without the leadership from a leader. Bragg (2000) suggests the use of the acronym "GREAT" or "Goals, Roles, Expectations, Abilities, Tools" as providing the fundamental foundation for team effectiveness and success. Gilpin (2002) describes an effective team that was driven with a common goal and a passion towards being successful through the term SoulPower. Heim (1996) reflects teamwork from the perspective of the "Zen Christian concept of selfless awareness" between teammates as followed by Phil Jackson in coaching a professional basketball team to many years of success (p. 654).

Bass et al. (1990) indicate that a correlation exists between the technical competence of the team leader and the effectiveness of the team. Additionally, knowledge familiarity with the job appears to be directly associated with effective performance and productivity (Goodman & Leyden, 1991; Guzzo et al., 1996). Other researchers feel that the potential of achieving a goal depends on the fit of the leader and the team members to perform their respective tasks (Bass, 1990). Selection of the right team members to enhance this fit increases the functioning of the team and its chances for success (Paris & Salas, 2000). In fact, some research advocates the concept of shared leadership as a construct that also determines team effectiveness (Pearce & Sims, 2000, 2002). Moment and Zaleznik (1963) find a strong correlation between the ability of the leader to interact and guide with the group s purposes and needs (p. 10). However, effectiveness can vary from team to team even when they are virtually identical (Hughes et al., 2002, p. 301).

Many researchers report that the level of support provided by an organization to its team affects attitudes of team members, productivity, and effective performance (Guzzo et al., 1996; Harrington-Mackin, 1994; Hyatt & Ruddy, 1997). This required support includes: "clear engaging direction, information, data, resources, rewards, and training" to increase team effectiveness and performance (Hyatt et al., 1997). Ogbonna and Harris (2000) recognize the importance of the organizational context and culture to maintaining a competitive advantage in a dynamic and complex business environment. Research supports the correlation of a team believing in success with actual success and effectiveness in the right climate (Pearce, Gallagher & Ensley, 2002).

Motivation of team members decreases when organizational support is lacking (Harrington-Mackin, 1994). Conversely, the more team building a leader can accomplish will increase the motivation of the team members to be more productive (Roche, 1994, p. 28). The type of work being performed, the interdependence of the team and team members, and the maturity and developmental level of the team also affect motivation (Janz & Colquitt, 1997). The team developmental cycle of forming, storming, norming, and performing helps explain the relationships and dynamics that establish team member attitudes and behaviors through the growth stages (Hughes et al., 2002). These stages are natural within the growth of a team and some teams progress through these stages without too much effort, but no team jumps from forming to performing without progressing through the other stages (Robbins & Finley, 1995).

The following suggested responsibilities of the team leader should result in enhancing performance:

- Transmits information, knowledge, and skills, in a timely fashion to team members
- Interprets and applies policies, work specifications, and job orders for the team
- Teaches team members how to manage work processes effectively and to evaluate results
- Builds communication channels between departments and eliminates duplication of effort
- Encourages team to identify what can be done differently or better

- Models proper team behavior in all areas; helps establish team climate and shape attitudes
- Promotes self-discipline in team members
- Encourages risk taking among team members by confronting group-think
- Supports goals of the team to internal and external customers
- Reinforces and rewards proper team behavior
- Troubleshoots for the team in areas of expertise
- Communicates team progress to management
- Serves as mediator during team conflicts to create win/win resolution
- Guides and shapes the direction toward a team culture (Harrington-Mackin, 1994, pp. 14-15)

Guzzo et al. (1996) identify three indicators of group effectiveness: group-produced outputs (quantity or quality, speed, customer satisfaction, and so on), the consequences a group has for its members, or the enhancement of a team's capability to perform effectively in the future (p. 309). Hughes et al. (2002) list eight characteristics, the first six associated with task accomplishment and the last two with group maintenance or interpersonal aspects of teams, that effective teams have in common (pp. 302-304):

Everyone on the team understands the mission and expects high performance standards

Leaders of effective teams recognize their equipment, training facilities and opportunities shortfalls and sought opportunities to strengthen them

Leaders assess the technical skills of team members on a

continuous basis

Leaders attempt to secure those resources and equipment that

help attain team effectiveness

Leaders spend a great deal of effort planning to optimize their

strengths and effectiveness

Leaders spend a great deal of effort organizing their resources

High levels of communications are pursued

Interpersonal conflicts are minimized through effective

communications

Gist et al. (1987) and Guzzo et al. (1996) found that the size of

team influences effectiveness and that larger organizations are more

productive than smaller organizations, but are less efficient. Team

member heterogeneity is positively correlated to the creativity and the

decision-making effectiveness of teams (Guzzo et al., 1996, p. 311).

The design of a team could be a factor in its effectiveness.

Specifically, four variables need to be present (Hughes et al., 2002, pp.

304-305):

Task structure : Is the task evident and is the structure consistent

with the capabilities of the team?

Group boundaries : Is the staffing of the team and the technical

and personal competencies of the members adequate for mission

accomplishment?

Norms : Has an outside organization or the team itself defined

specific standards succinctly so that expectations are understood?

Authority : Is the environment flexible and conducive for

empowerment of team members to allow for changing

requirements?

Hersey et al. (2001) find that effective leaders will not only employ

personal power, power of follower acceptance, but also general

supervision. Establishing a common vision with team members will result

in a shared commitment to a meaningful objective and can increase

energy levels and productivity to new heights (Bethel, 1990; Bradford &

Cohen, 1984). A study conducted by a professor of management at the

University of Nebraska concluded that effective managers spend most of

their time in communications and human resources management (Hersey

et al., 2001, p. 129).

Amason and Thompson (1995) and Lord (1977) identify the

management of conflict as yet another important dimension that impacts

team effectiveness. A certain level of conflict occurs during the storming

phase of the developmental cycle of the group, which is typical of group behavior (Hughes et al., 2002; Knowles & Shepherd, 1955). However, open communications and active involvement of team members in problem area discussions and conflict resolution reduces the impact of conflict within effective teams (Amason et al., 1995). Brown and Miller (2000) report a decrease in conflict behaviors with an increase in collaboration when stress situations increased. Devine and Clayton (1999) found that low levels of conflict were more frequently associated with effective teams. The occurrence of organizational dysfunction resulting in conflict was also more prevalent among teams that were resistant to change due to modifications in job assignments, workloads, and organizational support (Kirkman, 2000).

Avery (1999) suggests that a collaborative leader should communicate an equal emphasis on the "what" and the "why" of a particular task to increase the understanding and energize the motivation of team members towards goal accomplishment. The establishment of a shared culture in which there is no difference between "them" or "us" reflects the true ideal collective attitude of collaborative spirit (Barnett & McKowen, 1998). Fitz-Enz (1997) views, Teams are a visible manifestation of the transition toward collaboration as the keystone of business management (p. 3). Hays (1999) interprets collaboration as the wave of the future to success in a competitive business environment that advocates "do more with less," make quicker decisions, and implement

necessary change requiring higher interpersonal leadership skills and

process management abilities. Knowles et al. (1955) acknowledge the

shift of the leader s role from one who plans for, thinks for, takes

responsibility for, and directs other people, toward the notion that the

leader is primarily a convener, trainer and co-coordinator for a group (p.

12). Jack Welch of General Electric recommends an "informal" approach

to interpersonal interactions to enhance confidence levels and inspire a

climate of collaboration (Huey & Colvin, 1999). Directive leadership relying

on positional or legitimate power may be necessary when team members

must work in collaboration, but the team members are not unified and may

be new to the team, whereas participative leadership is more appropriate

when the team members are already trained, experienced, and committed

(Bass et al, 1990; Pearce & Sims, 2002). Bryman (1986) adds that

participative leadership increases the contentment of team members and

production when the members are allowed to participate in the decision

making process. Participative leadership is at its best when personal

growth is a goal and effectiveness depends on people taking initiative and

responsibility for their actions (Dimock, 1970a, p. 14). Lampe (1994) adds

that collaborative leadership requires both sensitivity to values differences

and negotiation prowess (p. 504).

New team members should strive to establish team credibility and trust as soon as possible in order to be accepted as full-fledged team members, eliminate any perceptions of being an outsider, and to re-establish a stable work climate (Benson-Armer & Stickel, 2000). Ilgen and Klein (1988) advocate that the organization attempt to positively influence the sense of belonging of new team members to increase individual performance and productivity, and that these efforts be monitored and changed as necessary to keep the process rewarding. Research supports that a strong group identity results in team stability, enhanced productivity, less personnel turnover, and ultimately success (McDonald & Hutchenson, 1998).

Careless and de Paola (2000) found that team cohesion is a key factor in models of effective work teams and should be viewed from the perspectives of interpersonal influence within the team and task cohesion towards task accomplishment. Guzzo et al. (1996) confirm the positive correlation between cohesion and productivity. Most of the time the cohesion found within a group experiencing growth is high, but sometimes a group may become too cohesive to allow newcomers to join, which will result in stagnation (Dimock, 1970b). An understanding of the natural growth of a group through the development cycle helps a leader explain and react to the various relationships of group cohesiveness and productivity found in the behavioral dynamics of the team (Hughes et al.,

2002; Knowles et al., 1955). In those instances that motivation is the

reason for the team cohesiveness, Lord (1977) recommends that the task

be completed as rapidly as possible to maximize the efficiencies.

The following sound advice is provided from the perspective of co-

leaders for execution of their duties and responsibilities and effective

interaction with their leader:

- Know thyself
- Know thy leader
- Avoid titanic clashes
- Give your bosses what they need, as well as what they want
- Find out what the enterprise needs most and deliver it superbly
- Don t sell your soul (or ruin your body)
- Lead as well as follow
- Know when to stay put
- Know when to walk away
- Define success on your terms
  (Heenan & Bennis, 1999, pp. 264-272)

Harrington-Mackin (1994) summarizes the most common

characteristics of productive teams (p. 21):

- Team goals are as important as individual goals; members are able to recognize when a personal agenda is interfering with the team s direction.
- The team understands the goals and is committed to achieving them; everyone is willing to shift responsibilities.
- The team climate is comfortable and informal; people feel empowered; individual competitiveness is inappropriate.
- Communication is spontaneous and shared among all members; diversity of opinions and ideas are encouraged.
- Respect, open-mindedness, and collaboration are high; members seek win/win solutions and build on each other s ideas.

- Trust replaces fear, and people feel comfortable taking risks; direct eye contact and spontaneous expression are present.
- Conflicts and differences of opinion are considered opportunities to explore new ideas; the emphasis is on finding common ground.
- The team works on improving itself constantly by examining its procedures, processes, and practices, and by experimenting with change.
- Leadership is rotated; no one person dominates.
- Decisions are made by consensus and have the acceptance and support of members.

Johnson (1996) and Labich (1992) agree with the above characteristics and reflect that they are essential guidelines to create a participative and collaborative environment that is focused and effective. The basis of collaborative leadership is creating a shared understanding of and commitment to the group s purpose and goal (Rawlings, 2000).

Leadership Ineffectiveness

Team members perceive external factors as hindering team effectiveness and contributing to the ineffectiveness of the team (Kline et al., 1996). The study of leadership effectiveness would be incomplete if some of the reasons for ineffectiveness were not addressed. Harrington-Mackin (1994) reflects the following as potential reasons for team ineffectiveness (p. 22):

- Its structure is incompatible with hierarchical organizational structure.
- It lacks visible support and commitment from top management.

- It has focused on task activities to the exclusion of work on team member relationships.
- Its members lack self-discipline and are unwilling to take responsibility for their own behavior and actions.
- The team has too many members and lacks the strong structure necessary to deal with a large team.
- Team members are unwilling to recognize and accept the patterns and stages of team process.
- The team has experienced poor leadership within and/or outside the team; there has been resistance from first-line supervisors.
- The organization has failed to use team efforts in any meaningful way.
- Members have received insufficient training.

Measuring Team Effectiveness

The variety of methods used to study team effectiveness is very broad and is oftentimes affected by constraints and limitations imposed by the environment being studied. Additionally, team metrics are oftentimes unique to the team being evaluated making interpretation and verification difficult (Fitz-Enz, 1997; Guzzo et al., 1996). One commonality among many of the approaches is that the analysis metric involved more than one measure. Combinations sometimes included comparisons of soft analysis of team perceptions with hard data such as production numbers. While the use of in-house surveys can be tailored specifically to the organization been evaluated, Henderson and Green (1997) express caution regarding their use due to the potential of not being scientifically valid and recommend the use of commercially available surveys that have been previously tested and validated.

The measurement of effectiveness is a performance variable that needs to be compared to other performance variables (Mahoney, 1988). Componation, Utley, and Swain (2001) identify six success measures of assessing team effectiveness, but conclude that due to the organizational context variance between teams that evaluating effectiveness from the standard measurement approach of quality, cost, and schedule is appropriate. Devine et al. (1999) confirm that a team s context may have an influence on a team s effectiveness. Fiedler (1958) adds that two concurrent attributes must be fulfilled prior to attempting to predict a leader s effectiveness: leader acceptance by the team members and leader psychological separation from the team members. These conditions seem to be important otherwise evaluations would not be able to differentiate the leader from the team members. Hays (1999) asserts that the measurement of success will depend on how effective a leader is in achieving collaboration towards goal accomplishment within a team.

Summary

Upon completion of an exhaustive literature review, it is evident that no single leadership theory or model is effective due to the many variables that influence each situation. In fact, at times a mixture of several approaches may be necessary to optimize the performance of the team to increase effectiveness and maximize productivity. Each team is unique and is different from one another. Its operational context, environment, personnel assigned, all create a very unique and special setting for a team leader to exhibit leadership under varying circumstances. Therefore, analysis of each situation and application of a single or hybrid mix of theories to best deal with a situation is constantly necessary and may make the difference between success and failure.

One does not become a leader through possession of a certain set of traits that create differentiation, but rather it is active participation in the collaborative relationships and associations with team members that elevates one to the leadership position. Then, the influence of variables and flux of situations challenges one s leadership. Demonstration of the ability to cooperatively handle tasks while constantly evaluating information, assessing competence, interpreting team dynamics, and sometimes relying on intuition to achieve task completion and goal accomplishment establishes a leader s reputation and success.

A leader plays an important role as a mediator between superiors and subordinates. Successful leaders have influence with superiors. Subordinates rely on the leader to be able to anticipate the expectations from above and provide a shield from the superior s wrath when necessary. A leader s effectiveness depends on his ability to get work accomplished through others. The ability to plan and then direct, delegate, negotiate, and participate through collaboration in the achievement of a goal will determine a leader s success and effectiveness. The survival of a team and the future of the team leader are dependent on the leadership exhibited by the leader to keep collaborative efforts moving toward a common goal and to maintain productivity at the expected level.

This research paper adopted the six key dimensions presented by Frank LaFasto and Carl Larson as essential elements to team leader effectiveness (LaFasto et al, 2001). The focus was to further substantiate these dimensions as necessary to achieve effectiveness and productivity and eventually success in the UXO industry through collaboration between the team leader and the team.

Chapter 3

Research Methodology

Introduction

This chapter will focus on the methods used in this research. It will discuss the approach, the data collection process, the analysis methods used to evaluate the data, and limitations of the study.

This research is based on observations and data collected during a specific case study of an actual UXO clearance project. Work teams are used extensively within this project. This study focuses entirely on the collaborative relationships of those teams as they execute duties assigned in the accomplishment of their tasks. Specifically, the ability of the team leader to focus on the goal, ensure a collaborative climate, build confidence, demonstrate sufficient technical know-how, set priorities, and manage performance is evaluated to determine why some teams are effective and able to meet their production goals while others cannot. The findings of this study will provide guidance for UXO industry leaders to use in order to maximize production.

Research Variables

Work teams comprising the entire production organization of a project were used as the focus of this study. These teams were not modified to accomplish this research. Consequently, various variables that are inherent to an intact team were not controlled. Team member and team leader skills, tenure within a team, personnel assignment to a team, and team size were examined and represent the composition variables.

Team members were assigned to a team based on a position vacancy and availability. Each team member was required to complete contractor unique training prior to commencement of work on the site. This training provided the requisite skills and knowledge for each team member to execute duties and responsibilities assigned. A demonstration of a team member s skill level was validated through a quality control test prior to commencement of work. Personnel were replaced as necessary due to vacancies or terminations of incumbents. Therefore, the tenure of each member of the team varied.

A team leader led each team. Each team leader, with the exception of one, was a graduate of Explosive Ordnance Disposal School and had served as an explosive ordnance disposal technician or unexploded ordnance specialist for a minimum of seven years. Their selection to the team leader position was based on position vacancy and availability, unexploded ordnance disposal experience as reflected on a resume, and performance in other positions within the project. Each team leader was required to complete contractor unique training and an unexploded ordnance refresher course with the attainment of a minimum score prior to commencement of work on the site. This training provided the requisite skills and knowledge for each leader to execute the technical duties and responsibilities assigned. Each team leader s technical skill level and command and control ability was also validated through a quality control test prior to assignment as a team leader. Leaders were replaced as necessary due to vacancies or terminations of incumbents. Therefore, the tenure and skill level of each team leader varied.

All teams work a four day, ten hour work shift. The average production team size varies based on the duties and responsibilities assigned to that team. The survey, data recording, and subsurface geophysical detection team consists of two people; assessment team and demolition teams, three; excavation team, seven; area preparation team, eight; range clearance team, seventeen; material management team from three to nine; and a quality control team from three to five personnel.

Team Task

Researchers stress the importance of taking into consideration the differences in the team task when evaluating effectiveness and productivity. This is important because the type of activities required by a team to be successful vary based on the task being performed. For example, the flow of information, communications, coordination, and collaboration should be easier and more effective within a small team versus a large one. In this study, the overall objective of the effort is unexploded ordnance clearance and the task assigned to each team varies based on their role within the effort, but the nature of the work and tasks is quite comparable.

Tasks assigned to a team are defined in standard operating procedures. Each member of the team is required to read and acknowledge understanding of the procedures. Safety and quality control personnel ensure compliance to the standard operating procedures during team surveillances and inspections. Within the overall clearance effort, tasks are accomplished sequentially one after the other. These tasks are mostly routine in nature and only vary due to environmental conditions, quality non-conformances, or logistical problems.

Research Items

      The purpose of this research was to study the relationship between the leadership demonstrated by the team leader and the effectiveness and productivity of the team. The following hypotheses were tested in this research through the use of two survey questionnaires developed by Frank LaFasto and Carl Larson (2001, pp. 151-154):

1. Clearly defining the team goal contributes to team effectiveness.
2. A collaborative climate increases productivity and team effectiveness.
3. Confidence of team members enhances performance, which results in better productivity and effectiveness.
4. Job knowledge contributes to increased performance.
5. Establishment of priorities helps achievement of goals.
6. Clear and recognized performance objectives enhance focus and better performance.

Team Effectiveness

      Multiple measures were used to evaluate the effectiveness of the teams and team leaders. Team member evaluations of the team leader, self-evaluations of the team leader, and manager performance ratings were used for this evaluation. Hard or objective production records were also used for most of the teams studied.

Different surveys were used to avoid common method variation influences to become a factor. The team member surveys were worded so that they were directed to assess the performance of the team leader. The team leader s version was directed towards a self-evaluation within the same areas as the team member surveys. The manager performance ratings were obtained through another survey that rank-ordered the teams.

Team Leader and Team Member Surveys

Based on the caution noted in the literature review referring to the establishment of new survey instruments, existing surveys applicable to the focus of the study were evaluated. The survey instrument adopted for this study, known as the Collaborative Team Leader Instrument, was developed by Frank LaFasto, Ph.D. and Carl Larson, Ph.D. from an evaluation of approximately 600 team leaders and measures team leader effectiveness across six key dimensions: focuses on the goal, ensures a collaborative climate, builds confidence, demonstrates sufficient technical know-how, sets priorities, and manages performance (LaFasto et al.,

2001, p. 96). A copy of the team leader and team member surveys used is included as Appendices A and B, respectively.

Two reliability tests were conducted to measure the internal reliabilities of the six dimensions using two different groups. The Cronbach alpha results were consistent and are shown in the table below.

Dimensions	Sample One	Sample Two
1. Focuses on the Goal	.92	.90
2. Ensures a Collaborative Climate	.94	.90
3. Builds Confidence	.90	.92
4. Demonstrates Technical Know-how	.90	.79
5. Sets Priorities	.92	.88
6. Manages Performance	.94	.94

Table 8 Internal Reliabilities for Six Leadership Dimensions
SOURCE: Frank M. J. LaFasto & Carl E. Larson, *When teams work best: 6,000 team members and leaders tell what it takes to succeed*, p. 210. Copyright © 2001 by Sage Publications, Inc. Reprinted by permission of Sage Publications, Inc.

Additionally, one-way analysis of variance conducted to differentiate between effective and ineffective leaders validated the instruments ability to discriminate between the team leaders (LaFasto et al., 2001).

Manager Performance Rating Survey

A copy of the survey developed for the managers is included as Appendix C. This survey focused on an effectiveness assessment of the teams under each manager s purview. The manager was required to provide an overall rating for each team relative to other teams within the project and to identify the three most important criteria for an effective team.

Data Collection

In the initial stages of this study, qualitative information from various interactions with the teams under study guided the focus of the research. These interactions established the foundation for an understanding of the importance of the survey and the need to provide an honest response to the items. Additionally, it was important to assure participants that participation or lack thereof would not result in termination of job or jeopardize future promotions.

Prior to administering the survey, project and union approvals were required. Project approvals included upper management concurrence and approval from the human resource representatives of the various companies involved with the project. Due to a perceived benefit from the findings and conclusions of the research by the project, no objections were expressed. The union was not as cooperative and wanted assurances that the survey was voluntary. This agreement to participate in a voluntary manner impacted survey participation. A copy of these requests and approvals are included as Appendix D.

Copies of the team leader and team member surveys were provided to each of the managers to survey each team. A briefing sheet explaining the survey and assuring participants of survey confidentiality was also included and was read prior to distributing the surveys. A copy of the briefing sheet is included as Appendix E. Surveys were administered during convenient times when a team was not actively involved in the field. Most participants completed the survey within 15 to 20 minutes. Surveys were then collected by the manager and returned to the researcher. Managers completed the manager s performance rating survey at a time of their convenience and returned it to the researcher.

Analysis

Scale reliability analysis, aggregation analysis, and bivariate correlation were used to analyze each data set of this study. Additionally, response rates were calculated for each team participating in the survey. Also, a comparison of a team identified as an effective team was made to actual production rates to validate the findings that a correlation exists between effectiveness and production.

The scale reliability analysis was conducted using Cronbach s alpha. Reliability refers to the consistency of the measure of an item to the overall response. This measure is important because it assesses the level of consistency between survey items. Reliability is the variability in the responses by the respondents instead of the scaling of the differences of survey items. A level of internal reliability that is acceptable is a Cronbach s alpha greater than 0.7. This test was applied to all the survey responses and was used to evaluate the scale reliabilities of the team leader and team member survey instruments to this criterion.

Bivariate correlation using Pearson product moment correlation was used to measure the linearity ranging from 1 to 1 between two variables. Scatter graphs of each survey item for each pair of team leader and team member responses were reviewed. This examination focused on identifying potential relationships between the variables. The bivariate correlation analysis for each pairing focused on aggregated team means to normalize the varying number of respondents within each team and category. Additionally, p-values were checked to determine significance of correlation.

Database of Study

Obtaining a representative sample of response rates is always a concern when conducting research with a survey instrument. In this study, individuals are providing the raw data responses, but the focus was the team; specifically, the relationship between the team members and the team leader. While all team personnel were afforded the opportunity to participate in the study, the labor union insistence that the survey instruments be voluntary in nature reduced the response rate obtained. Additionally, some of the team leaders were reluctant to participate for various reasons. Since a random sampling of each team was not pursued, much of the data provided could not be used since the minimum requirement to conduct a comparison was a team leader and at least one team member, and in some instances this was not obtained. However, all the data was used to calculate Cronbach s alpha value and response rate analysis.

The total number of teams involved in this survey was 28. The total number of teams in which both a team leader and at least one team member responded was 16. The total number of personnel involved in this survey was 152 of which 19 were team leaders and 133 team members. The average response rate was 68% for the team leaders and 86% for team members. Average tenure in team was 14 months for team leaders and 18 months for team members. Response rates by team, team size, and average time with the team are summarized in Table 9 for team leaders and Table 10 for team members.

Team	Response Rate Percentage	Size	Average Tenure in Team (months)
1	100	1	16
2	100	1	5
3	100	1	19
4	100	1	23
5	100	1	12
6	100	1	11
7	100	1	8
8	100	1	8
9	100	1	19
10	100	1	14
11	100	1	11
12	100	1	13
13	100	1	27
14	100	1	21
15	100	1	15
16	100	1	20
17	100	1	5
18	100	1	18
19	100	1	7

Table 9 Team Leader Survey Response Rates

83

Team	Response Rate Percentage	Size	Average Tenure in Team (months)
1	93	14	15
2	60	9	5
3	93	13	9
4	100	7	8
5	71	5	37
6	100	7	26
7	100	7	15
8	86	6	42
9	43	3	17
10	29	2	32
11	43	3	21
12	29	2	27
13	14	1	30
14	100	7	21
15	100	7	18
16	57	4	18
17	29	2	18
18	86	6	14
19	100	7	16
20	71	5	5
21	50	1	5
22	63	5	6
23	100	4	7
24	55	6	12

Table 10 Team Member Survey Response Rates

Validity and Originality of the Data

This case study was unique because it focused on original data obtained during an actual, active UXO clearance project. Techniques used to obtain the data involved the administering of survey questionnaires in the field where the workers executed their duties and production records depicting actual production rates.

Valid and reliable information was obtained through a controlled environment in which respondents were monitored to ensure responses were in fact their own and were not collaborated. Survey questionnaire administrators remained with the team until surveys were completed. This controlled approach in the administration of the survey ensured unbiased and objective results. Surveys were then returned to the researcher for data compilation and analysis.

Data obtained during this study allowed the researcher to conduct an analysis and interpret the results from an empirical basis. Findings of fact, opinions, and conclusions are supported from well-reasoned and objective evidence.

Limitations of the Data

Many variables were involved in this study. The variations in the environmental conditions encountered by the work teams as they proceeded through the various phases of UXO clearance affected production levels. Previous leadership instruction and experiences of the team leaders were factors as well. Personalities, cultures, and individual differences of the various team personnel came into play. Level of involvement of the leaders with their teams impacted attitudes, motivations, and work behaviors. Diversity in the nature of the tasks can be a challenge to any analysis and will not be evaluated in this study. All the teams were viewed as performing similar UXO operations without task differentiation.

Some of the limitations of this study impacted its conclusions. Studying only one, very unique project presented a item as to the validity of the comparison of the results to similar projects. The size of the sample used in the research was rather small and may not be statistically significant. And lastly, the development of surveys as a measurement tool and interpretation of outcomes were subject to biases by the interpreter.

Summary

The analysis of this case study was straightforward, but was hampered by a small sampling size that may cause some bias in the data and affect the validity of some of the conclusions. Unfortunately, not being able to enforce 100% participation of team members and team leaders because of the influence of labor unions, this result was unavoidable. However, the results are viewed as favorable as many good concepts and theories have been addressed that could still be adopted or have a positive influence on enhancing team leader performance and effectiveness within the UXO industry.

Chapter 4

Data Analysis

This chapter provides a summary of the results of the analyses described in Chapter 3. Results will be summarized separately for the team leader and team member data sets. For clarity of presentation for the various analyses, a system of abbreviations is used to identify each specific key dimension as identified by LaFasto et al. (2001, p. 96) to the corresponding survey items. When necessary to differentiate between the team leader and the team member, an "L" is used prior to the abbreviation to specify that the abbreviation is used in the context of a team leader. The specific survey item in the questionnaire is identified with a number following each abbreviation, which is consistent with the actual numbering of each item.

Table 11 Dimension Abbreviations

Dimension	Abbreviation
Focus on the goal	G
Ensure a collaborative climate	CC
Build confidence	C
Demonstrate sufficient technical know-how	T
Set priorities	P
Manage performance	MP

Scale Reliability Analysis

The scale reliability analysis for each survey item was assessed using Cronbach's alpha. The reliability coefficients are summarized in Table 12 for both the team leader and team member data sets. In addition, the number of survey items for each dimension is depicted. Reliabilities were determined based on the entire set of respondents (N= 20 for team leader data and N= 134 for team member data).

Table 12 Internal Reliabilities for each Dimension

Dimension Abbreviation	Team Leader (N = 20)		Team Member (N = 134)	
	Reliability	Number of Items	Reliability	Number of Items
G	0.739	6	0.997	6
CC	0.6678	10	0.998	10
C	0.6969	7	0.997	7
T	-0.0613	5	0.998	5
P	0.5126	5	0.997	5
MP	0.6154	7	0.996	7

Reliabilities for all dimensions for the team members were not only greater than 0.7, the rule-of-thumb acceptable level of agreement, but all were above 0.99. This indicated that all the survey items were reliable indicators of each dimension. These results were consistent with those obtained by two initial surveys conducted by LaFasto et al. (2001).

However, the Cronbach alpha reliabilities for the team leaders were not as consistent as the results for the team members. In fact, it appears that the internal reliabilities reported by LaFasto et al. (2001) were for two samples of team member personnel and not team leaders. The reliability of three of the dimensions (focus on the goal, ensure a collaborative climate, and build confidence) was found to be acceptable when rounded up. The other three (demonstrate sufficient technical know-how, set priorities, and manage performance) were determined not to be reliable indicators of these dimensions. It is interesting to note that the same two dimensions (demonstrate sufficient technical know-how and set priorities) were also found to be less reliable than the other dimensions during the two samples conducted by LaFasto et al. (2001) and depicted in Table 8.

After closer review of the survey items within each of the three-team leader dimensions that resulted in an internal reliability score of less than 0.70, the following explanations were developed for each dimension:

1. Demonstrate Sufficient Technical Know-how Dimension

   - The Cronbach's alpha value for Item LT24 was -0.0654 indicating a negative reliability. A review of the raw data and the mean values reflected 19 of 20 respondents reporting a 4.0 and 1 of 20 reporting a 3.0 out of a possible 4.0 for all 20 respondents with a mean of 3.9375 for Item LT24. It is concluded that the small sample size may have adversely affected the results of the reliability test for this item.

- The Cronbach's alpha value for Item LT25 was -0.4118, the most negative of all of the results within the entire data set within this dimension, indicating a negative reliability. A review of the raw data reflected 18 of 20 respondents reporting a 4.0 and 2 of 20 reporting a 3.0 out of a possible 4.0 for all 20 respondents with a mean of 3.9375 for Item LT25. The Cronbach s alpha value for this dimension data set was recalculated without LT25. The score for the entire set actually dropped further to -0.4118. This indicates that the LT25 variable is correlated with the other items in the dimension data set. It is concluded that the small sample size may have adversely affected the results of the reliability test for the LT25 item.

- The Cronbach's alpha value for Item LT26 was -0.0152 indicating a negative reliability. A review of the raw data and the mean values reflected 19 of 20 respondents reporting a 4.0 and 1 of 20 reporting a 3.0 out of a possible 4.0 for all 20 respondents with a mean of 3.9375 for Item LT26. It is concluded that the small sample size may have adversely affected the results of the reliability test for the LT26 item.

- The Cronbach's alpha value for Item LT27 was -0.2302, the second most negative of all of the results within the entire data set within this dimension, indicating a negative reliability. A review of the raw data reflected 17 of 20 respondents reporting a 4.0 and 3 of 20 reporting a 3.0 out of a possible 4.0 for all 20 respondents with a mean of 3.875 for Item LT27. The Cronbach s alpha value for this dimension data set was recalculated without LT27. The score for the entire set actually remained the same at -0.2302. This indicates that the LT27 variable is not a factor to the other items in the dimension data set. It is concluded that the small sample size may have adversely affected the results of the reliability test for the LT27 item.

- The Cronbach's alpha value for Item LT28 was 0.3743, less than the acceptable level of agreement, indicating a weak reliability. A review of the raw data reflected 13 of 20 respondents reporting a 4.0 and 7 of 20 reporting a 3.0 out of a possible 4.0 for all 20 respondents with a mean of 3.6875 for item LT28. It is concluded that the small sample size may have adversely affected the results of the reliability test for the LT28 item.

2. Set Priorities Dimension

- The Cronbach's alpha value for item LP29 was 0.4930, less than the acceptable level of agreement, indicating a weak reliability. A review of the raw data reflected 17 of 20 respondents reporting a 4.0 and 3 of 20 reporting a 3.0 out of a possible 4.0 for all 20 respondents with a mean of 3.9375 for item LP29. It is concluded that the small sample size may have adversely affected the results of the reliability test for the LT29 item.

- The Cronbach's alpha value for item LP30 was 0.4169, the second lowest for this dimension data set and less than the acceptable level of agreement, indicating a weak reliability. A review of the raw data reflected 12 of 20 respondents reporting a 4.0 and 8 of 20 reporting a 3.0 out of a possible 4.0 for all 20 respondents with a mean of 3.5625 for Item LP30. Cronbach s alpha value for this dimension data set was recalculated without LP30. The score for the entire set actually remained the same at 0.4169. This indicates that the LP30 variable is not a factor to the other items in the dimension data set. It is concluded that the small sample size may have adversely affected the results of the reliability test for the LP30 item.

- The Cronbach's alpha value for item LP31 was 0.2832, the lowest within the dimension data set and less than the acceptable level of agreement, indicating a weak reliability. A review of the raw data reflected 14 of 20 respondents reporting a 4.0 and 6 of 20 reporting a 3.0 out of a possible 4.0 for all 20 respondents with a mean of 3.625 for item LP31. Cronbach s alpha value for this dimension data set was recalculated without LP31. The score for the entire set actually dropped further to 0.2832. This indicates that the LP31 variable is correlated with the other items in the dimension data set. It is concluded that the small sample size may have adversely affected the results of the reliability test for the LP31 item.

- The Cronbach's alpha value for item LP32 was 0.4837, less than the acceptable level of agreement, indicating a weak reliability. A review of the raw data reflected 14 of 20 respondents reporting a 4.0 and 6 of 20 reporting a 3.0 out of a possible 4.0 for all 20 respondents with a mean of 3.6875 for Item LP32. It is concluded that the small sample size may have adversely affected the results of the reliability test for the LT32 item.

- The Cronbach's alpha value for item LP33 was 0.5424, less than the acceptable level of agreement, indicating a weak reliability. A review of the raw data reflected 18 of 20 respondents reporting a 4.0 and 2 of 20 reporting a 3.0 out of a possible 4.0 for all 20 respondents with a mean of 3.875 for item LP33. It is concluded that the small sample size may have adversely affected the results of the reliability test for the LP33 item.

3. Manage Performance Dimension

- The Cronbach's alpha value for item LMP34 was 0.6242, less than the acceptable level of agreement, indicating a weak reliability. A review of the raw data reflected 19 of 20 respondents reporting a 4.0 and 1 of 20 reporting a 3.0 out of a possible 4.0 for all 20 respondents with a mean of 3.9375 for Item LMP34. It is concluded that the small sample size may have adversely affected the results of the reliability test for the LMP34 item.

- The Cronbach's alpha value for item LMP35 was 0.5444, less than the acceptable level of agreement, indicating a weak reliability. A review of the raw data reflected 14 of 20 respondents reporting a 4.0 and 6 of 20 reporting a 3.0 out of a possible 4.0 for all 20 respondents with a mean of 3.625 for item LMP35. It is concluded that the small sample size may have adversely affected the results of the reliability test for the LMP35 item.

- The Cronbach's alpha value for item LMP36 was 0.5072, the lowest within the dimension data set and less than the acceptable level of agreement, indicating a weak reliability. A review of the raw data reflected 10 of 20 respondents reporting a 4.0, 9 of 20 reporting a 3.0, and 1 of 20 reporting a 2.0 out of a possible 4.0 for all 20 respondents with a mean of 3.4375 for Item LMP36. Cronbach s alpha value for this dimension data set was recalculated without LMP36. The score for the entire set actually remained the same at 0.5072. This indicates that the LMP36 variable is not a factor to the other items in the dimension data set. It is concluded that the small sample size may have adversely affected the results of the reliability test for the LMP36 item.

- The Cronbach's alpha value for item LMP37 was 0.5949, less than the acceptable level of agreement, indicating a weak reliability. A review of the raw data reflected 15 of 20 respondents reporting a 4.0, 3 of 20 reporting a 3.0, and 2 of 20 reporting a 2.0 out of a possible 4.0 for all 20 respondents with a mean of 3.6875 for item LMP37. It is concluded that the small sample size may have adversely affected the results of the reliability test for the LMP37 item.

- The Cronbach's alpha value for item LMP38 was 0.6321, less than the acceptable level of agreement, indicating a weak reliability. A review of the raw data reflected 14 of 20 respondents reporting a 4.0 and 6 of 20 reporting a 3.0 out of a possible 4.0 for all 20 respondents with a mean of 3.6875 for item LMP38. It is concluded that the small sample size may have adversely affected the results of the reliability test for the LMP38 item.

- The Cronbach's alpha value for item LMP39 was 0.5972, less than the acceptable level of agreement, indicating a weak reliability. A review of the raw data reflected 17 of 20 respondents reporting a 4.0 and 3 of 20 reporting a 3.0 out of a possible 4.0 for all 20 respondents with a mean of 3.8125 for item LMP39. It is concluded that the small sample size may have adversely affected the results of the reliability test for the LMP39 item.

- The Cronbach's alpha value for item LMP40 was 0.5250, the second lowest within the dimension data set and less than the acceptable level of agreement, indicating a weak reliability. A review of the raw data reflected 18 of 20 respondents reporting a 4.0 and 2 of 20 reporting a 3.0 out of a possible 4.0 for all 20 respondents with a mean of 3.9375 for item LMP40. Cronbach s alpha value for this dimension data set was recalculated without LMP40. The score for the entire set actually dropped further to 0.5250. This indicates that the LMP40 variable is correlated with the other items in the dimension data set. It is concluded that the small sample size may have adversely affected the results of the reliability test for the LMP40 item.

Aggregation Analysis

One-way analysis of variance was used to assess whether group level differences were significant for the survey responses within each dimension. These results are summarized in Table 13. Group level differences were not significant (p less than or equal to .05) for all

dimensions.

Table 13 One-way Analysis of Variance Results

Dimension	Sum of Squares	df	Mean Square	F	p
Focus on the goal	15.8663	42	0.3782	-4.26e13	-1.0
Ensure a collaborative climate	9.1591	42	0.2181	0	0
Build confidence	8.5455	42	0.2034	3.818e13	<0.0001
Demonstrate sufficient technical know-how	13.1591	42	0.3133	-1.76e14	-1.0
Set priorities	10.25	42	0.2441	0	0
Manage performance	8.5455	42	0.2035	-1.15e14	-1.0

Pearson Product Moment Correlation

Bivariate correlation using Pearson product moment correlation

was used to measure the linearity ranging from   1 to 1 between the team

leaders responses and team member responses to each survey item

within the key dimensions. Scatter graphs of each survey item for each

pair of team leader and team member responses were reviewed. This

examination focused on identifying potential relationships between the

variables. The bivariate correlation analysis for each pairing focused on

aggregated team means to normalize the varying number of respondents within each team and category. Additionally, p-values were checked to determine significance of correlation.

Focus on the Goal Dimension

A total of six pairs of team leader and team member responses were compared within this dimension. The correlation levels and significance probabilities for each item are addressed. Additionally, remarks are presented where needed to further clarify a finding.

1. LG1 and G1
   - Correlation between variables was positive at 0.3323 with a significance probability of 0.2086.
   - The mean of the responses for LG1 and G1 was 3.6875 and 3.7563, respectively.
   - The correlation result reflected agreement that the team leaders clearly defined the team goal to team members.

2. LG2 and G2
   - Correlation between variables was positive at 0.2694 with a significance probability of 0.3130.
   - The mean of the responses for LG2 and G2 was 3.5625 and 3.7187, respectively.

- The correlation result reflected agreement that the team leaders articulated the team goal in an inspirational manner.

3. LG3 and G3

  - Correlation between variables was positive at 0.3328 with a significance probability of 0.2078.

  - The mean of the responses for LG3 and G3 was 3.75 and 3.8256, respectively.

  - The correlation result reflected agreement that the team leaders did not allow politics to distract the teams from its goals.

4. LG4 and G4

  - Correlation between variables was positive at 0.1996 with a significance probability of 0.4586.

  - The mean of the responses for LG4 and G4 was 3.9375 and 3.6737, respectively.

  - The correlation result reflected agreement that the team leaders helped alignment of team member roles and responsibilities with team goals.

5. LG5 and G5

  - Correlation between variables was positive at 0.2758 with a significance probability of 0.3012.

- The mean of the responses for LG5 and G5 was 3.3125 and 3.5206, respectively.

- The correlation result reflected agreement that the team leaders reinforced the team goal.

6. LG6 and G6

- Correlation between variables was negative at -0.2730 with a significance probability of 0.3062.

- The mean of the responses for LG6 and G6 was 3.6875 and 3.7137, respectively.

- The correlation result reflected disagreement between the team leaders' and team members' perceptions. The team members viewed the team leaders as not adequately informing the team when changes to the team's goals were necessary, whereas the team leaders thought that the explanation was adequate.

Ensure a Collaborative Climate Dimension

A total of ten pairs of team leader and team member responses were compared within this dimension. The correlation levels and significance probabilities for each item are addressed. Additionally, remarks are presented where needed to further clarify a finding.

1. LCC7 and CC7

    - Correlation between variables was positive at 0.5789 with a significance probability of 0.0188.

    - The mean of the responses for LCC7 and CC7 was 3.875 and 3.8383, respectively.

    - The correlation result reflected agreement that the team leaders established a climate that was conducive to open communications.

2. LCC8 and CC8

    - Correlation between variables was positive at 0.5661 with a significance probability of 0.0223.

    - The mean of the responses for LCC8 and CC8 was 3.8125 and 3.7031, respectively.

    - The correlation result reflected agreement that the team leaders communicated openly and honestly with the team.

3. LCC9 and CC9

    - Correlation between variables was positive at 0.3921 with a significance probability of 0.1331.

    - The mean of the responses for LCC9 and CC9 was 3.6875 and 3.7556, respectively.

- The correlation result reflected agreement that the team leaders were not hesitant to discuss any issues with the team.

4. LCC10 and CC10

   - Correlation between variables was negative at -0.0638 with a significance probability of 0.8143.

   - The mean of the responses for LCC10 and CC10 was 3.625 and 3.6669, respectively.

   - The correlation result reflected disagreement between the team leaders' and team members' perceptions. The team members viewed some chronic problems existing that the team leaders were unable to resolve, whereas the team leaders thought that no chronic problems existed.

5. LCC11 and CC11

   - Correlation between variables was positive at 0.5060 with a significance probability of 0.0455.

   - The mean of the responses for LCC11 and CC11 was 3.3125 and 3.5206, respectively.

   - The correlation result reflected agreement that the team leaders did not tolerate non-collaborative conduct within the team.

6. LCC12 and CC12

- Correlation between variables was negative at -0.1692 with a significance probability of 0.5311.

- The mean of the responses for LCC12 and CC12 was 3.8125 and 3.7525, respectively.

- The correlation result reflected disagreement between the team leaders' and team members' perceptions. The team members viewed the team leaders as not recognizing and rewarding behavior that inspired a supportive and open team climate, whereas the team leaders thought that acknowledgements and rewards existed.

7. LCC13 and CC13

- Correlation between variables was positive at 0.1276 with a significance probability of 0.6378.

- The mean of the responses for LCC13 and CC13 was 3.8125 and 3.72, respectively.

- The correlation result reflected agreement that the team leaders created a work environment that promoted productive problem solving within the teams.

8. LCC14 and CC14

- Correlation between variables was negative at -0.1458 with a significance probability of 0.5902.

- The mean of the responses for LCC14 and CC14 was 3.4375 and 3.4494, respectively.

- The correlation result reflected disagreement between the team leaders' and team members' perceptions. The team members viewed the team leaders as allowing organization structure, systems, and processes to interfere with the accomplishment of team goals, whereas the team leaders thought that team goals were not interfered with by outside influences.

9. LCC15 and CC15

- Correlation between variables was positive at 0.0490 with a significance probability of 0.8569.

- The mean of the responses for LCC15 and CC15 was 3.625 and 3.67, respectively.

- The correlation result reflected agreement that the team leaders managed their personal control needs.

10. LCC16 and CC16

- Correlation between variables was positive at 0.4317 with a significance probability of 0.0950.

- The mean of the responses for LCC16 and CC16 was 3.625 and 3.6413, respectively.

- The correlation result reflected agreement that the team leaders did not allow their egos to interfere with the team.

Build Confidence Dimension

A total of seven pairs of team leader and team member responses were compared within this dimension. The correlation levels and significance probabilities for each item are addressed. Additionally, remarks are presented where needed to further clarify a finding.

1. LC17 and C17

   - Correlation between variables was negative at -0.2245 with a significance probability of 0.4031.

   - The mean of the responses for LC17 and C17 was 3.875 and 3.9331, respectively.

   - The correlation result reflected disagreement between the team leaders' and team members' perceptions. The team members viewed the team leaders as not ensuring that the team achieved results, whereas the team leaders thought that they were focused on result accomplishment.

2. LC18 and C18

- Correlation between variables was positive at 0.6011 with a significance probability of 0.0138.

- The mean of the responses for LC18 and C18 was 3.875 and 3.6306, respectively.

- The correlation result reflected agreement that the team leaders helped strengthen the self-confidence of team members.

3. LC19 and C19

- Correlation between variables was positive at 0.4553 with a significance probability of 0.0764.

- The mean of the responses for LC19 and C19 was 3.7718 and 3.875, respectively.

- The correlation result reflected agreement that the team leaders ensured that team members understood critical issues and important facts.

4. LC20 and C20

- Correlation between variables was positive at 0.6071 with a significance probability of 0.0126.

- The mean of the responses for LC20 and C20 was 3.9375 and 3.8087, respectively.

- The correlation result reflected agreement that the team leaders exhibited trust in the team members through assignment of meaningful levels of responsibilities.

5. LC21 and C21

   - Correlation between variables was positive at 0.1135 with a significance probability of 0.6755.

   - The mean of the responses for LC21 and C21 was 3.75 and 3.7718, respectively.

   - The correlation result reflected agreement that the team leaders were fair and impartial toward all team members.

6. LC22 and C22

   - Correlation between variables was positive at 0.2154 with a significance probability of 0.4231.

   - The mean of the responses for LC22 and C22 was 3.625 and 3.7712, respectively.

   - The correlation result reflected agreement that the team leaders were optimistic and focused on opportunities.

7. LC23 and C23

   - Correlation between variables was positive at 0.1377 with a significance probability of 0.6110.

- The mean of the responses for LC23 and C23 was 3.9375 and 3.6825, respectively.
- The correlation result reflected agreement that the team leaders looked for and acknowledged team member contributions.

Demonstrate Sufficient Technical Know-how Dimension

A total of five pairs of team leader and team member responses were compared within this dimension. The correlation levels and significance probabilities for each item are addressed. Additionally, remarks are presented where needed to further clarify a finding.

1. LT24 and T24

    - Correlation between variables was positive at 0.0481 with a significance probability of 0.8597.
    - The mean of the responses for LT24 and T24 was 3.9375 and 3.7656, respectively.
    - The correlation result reflected agreement that the team leaders understood technical issues involved in the achievement of team goals.

2. LT25 and T25

- Correlation between variables was negative at -0.1698 with a significance probability of 0.5296.

- The mean of the responses for LT25 and T25 was 3.9375 and 3.8431, respectively.

- The correlation result reflected disagreement between the team leaders' and team members' perceptions. The team members viewed the team leaders as not having sufficient experience with the technical issues affecting the team goals, whereas the team leaders thought that they had the requisite technical experience to accomplish the goal.

3. LT26 and T26

   - Correlation between variables was negative at -0.0793 with a significance probability of 0.7704.

   - The mean of the responses for LT26 and T26 was 3.9375 and 3.7644, respectively.

   - The correlation result reflected disagreement between the team leaders' and team members' perceptions. The team members viewed the team leaders as not receptive to technical advice from more knowledgeable team members, whereas the team leaders thought that they were receptive to technical advice.

4. LT27 and T27

- Correlation between variables was negative at -0.3062 with a significance probability of 0.2487.

- The mean of the responses for LT27 and T27 was 3.875 and 3.8688, respectively.

- The correlation result reflected disagreement between the team leaders' and team members' perceptions. The team members viewed the team leaders as incapable of assisting the team in analyzing complex issues affecting the teams goals, whereas the team leaders thought that they were capable in helping the team with complex issues.

5. LT28 and T28

- Correlation between variables was positive at 0.1566 with a significance probability of 0.5624.

- The mean of the responses for LT28 and T28 was 3.6875 and 3.7831, respectively.

- The correlation result reflected agreement that people outside the team viewed the team leaders as credible and knowledgeable.

Set Priorities Dimension

A total of five pairs of team leader and team member responses were compared within this dimension. The correlation levels and significance probabilities for each item are addressed. Additionally, remarks are presented where needed to further clarify a finding.

1. LP29 and P29

   - Correlation between variables was positive at 0.4372 with a significance probability of 0.0904.

   - The mean of the responses for LP29 and P29 was 3.9375 and 3.8081, respectively.

   - The correlation result reflected agreement that the team leaders focused on a manageable set of goals that would lead to goal accomplishment.

2. LP30 and P30

   - Correlation between variables was positive at 0.4571 with a significance probability of 0.0751.

   - The mean of the responses for LP30 and P30 was 3.7238 and 3.5625, respectively.

   - The correlation result reflected agreement between the team leaders and team members on the top priorities for team goal achievement.

3. LP31 and P31

- Correlation between variables was positive at 0.4467 with a significance probability of 0.0828.

- The mean of the responses for LP31 and P31 was 3.625 and 3.7769, respectively.

- The correlation result reflected agreement that the team leaders communicated and reinforced a focus on priorities to team members.

4. LP32 and P32

- Correlation between variables was negative at -0.0477 with a significance probability of 0.8606.

- The mean of the responses for LP32 and P32 was 3.6875 and 3.6769, respectively.

- The correlation result reflected disagreement between the team leaders' and team members' perceptions. The team members viewed the team leaders as diluting the efforts with too many priorities, whereas the team leaders thought that they were not diluting the priorities.

5. LP33 and P33

- Correlation between variables was positive at 0.0090 with a significance probability of 0.9736.

- The mean of the responses for LP33 and P33 was 3.875 and 3.7569, respectively.
- The correlation result reflected agreement that the team leaders explained reasons for changes to priorities to team members.

Manage Performance Dimension

A total of seven pairs of team leader and team member responses were compared within this dimension. The correlation levels and significance probabilities for each item are addressed. Additionally, remarks are presented where needed to further clarify a finding.

1. LMP34 and MP34
   - Correlation between variables was positive at 0.1353 with a significance probability of 0.6172.
   - The mean of the responses for LMP34 and MP34 was 3.9375 and 3.7713, respectively.
   - The correlation result reflected agreement that the team leaders explained performance expectations clearly to team members.
2. LMP35 and MP35

- Correlation between variables was positive at 0.0160 with a significance probability of 0.9530.

- The mean of the responses for LMP35 and MP35 was 3.625 and 3.67, respectively.

- The correlation result reflected agreement that the team leaders encouraged the team to establish an agreed upon set of values to guide the team.

3. LMP36 and MP36

- Correlation between variables was positive at 0.2761 with a significance probability of 0.3006.

- The mean of the responses for LMP36 and MP36 was 3.4375 and 3.5769, respectively.

- The correlation result reflected agreement that the team leaders ensured that rewards and incentives were consistent with achievement of team goals.

4. LMP37 and MP37

- Correlation between variables was negative at -0.0406 with a significance probability of 0.8814.

- The mean of the responses for LMP37 and MP37 was 3.6875 and 3.7163, respectively.

- The correlation result reflected disagreement between the team leaders' and team members' perceptions. The team members viewed the team leaders as not assessing the collaborative skills of team members and the results they achieved.

5. LMP38 and MP38

- Correlation between variables was positive at 0.5482 with a significance probability of 0.0279.

- The mean of the responses for LMP38 and MP38 was 3.6875 and 3.5506, respectively.

- The correlation result reflected agreement that the team leaders provided meaningful and developmental feedback to team members.

6. LMP39 and MP39

- Correlation between variables was negative at -0.2462 with a significance probability of 0.3580.

- The mean of the responses for LMP39 and MP39 was 3.8125 and 3.6275, respectively.

- The correlation result reflected disagreement between the team leaders' and team members' perceptions. The team members viewed the team leaders as not confronting and resolving issues when team member performance was inadequate.

7. LMP40 and MP40

- Correlation between variables was negative at -0.1950 with a significance probability of 0.4693.

- The mean of the responses for LMP40 and MP40 was 3.9375 and 3.625, respectively.

- The correlation result reflected disagreement between the team leaders' and team members' perceptions. The team members viewed the team leaders as not recognizing and rewarding team members superior performance.

Open Item Analysis

Two open-ended items were included in the team leader and team member surveys. The first item attempted to identify the strengths of the team leader and the second, the one or two changes that are most likely to improve the effectiveness of the team leader.

The team leaders responses to the first item regarding strengths included the following top ten in order of priority:

- Interpersonal skills
- Technical knowledge

- Leadership skills

- Open communications

- Integrity

- Motivation skills

- Organizational skills

- Perseverance

- Realistic goals

- Problem-solving ability

Each one of the above team leaders responses was related to one or more of the six key dimensions. This relationship will be discussed in the order of the above listing in the following sections.

The interpersonal skills response is related to the focus on the goal and ensure a collaborative climate dimensions. Both of these dimensions reflected significant positive correlation in the survey with the focus on the goal dimension resulting in five out of six items and the collaborative climate dimension resulting in seven out of ten items. Certainly a common element to both of these dimensions is clarity of communications. The team leaders felt that this skill, communicating clearly with team members, enhanced their relationships with each person and fostered a collaborative climate. This climate was further strengthened through a common focus that bonded the team into a unit with a common purpose.

The technical knowledge response coincides exactly with the demonstrate sufficient technical know-how dimension. This dimension did not fair as well in the survey as three of the five items reflected negative correlation indicating an inconsistent perception between team leaders and team members. All the team leaders prided themselves in their knowledge of ordnance and clearance procedures. This trait is not uncommon to professions in which life and death situations are commonplace.

The leadership skills response represents all six of the key dimensions. These dimensions represent competencies necessary for effective leaders. It was to be expected that the team leaders would view themselves as being effective leaders.

The open communications response is key to the collaborative climate dimension. Without freedom of speech without repercussions collaboration would be non-existent. From the team leaders perspectives, it was understandable that their perspectives were that a collaborative climate existed. Seven of the ten survey items with positive correlation confirmed this fact.

The integrity, motivations skills, and organizational skills responses are all strengths associated with building confidence and ensuring a collaborative climate. Six of seven and seven of ten survey items with positive correlation within the building confidence and collaborative climate dimensions supported these strengths. Without these strengths, trust would not occur and individuals would become disenchanted resulting in a disjointed organization. Certainly collaboration would suffer. Again it is interesting to note that the team leaders viewed themselves as providing these key ingredients to building a collaborative climate that exuded confidence.

The responses perseverance and realistic goals were consistent with the focus on the goal dimension. Six out of seven survey items with a positive correlation within this dimension confirmed this strength. Not only did the team leaders demonstrate an insistence that the goal was achievable, but they ensured that it was within reach and pursued it with vigor.

The last strength on the list, problem-solving ability, is an important response that demonstrates a need to create a collaborative climate while maintaining focus on the goal. Seven out of ten of the survey items with a positive correlation supported this perception. This strength was not only inherent in the team leaders abilities, but also within the team as a whole as they pursued goal accomplishment with an understanding that decisions affected success and the goal inspired teamwork.

The team leaders responses to the second item regarding changes included the following in order of the one most commonly selected:

- Additional training in leadership skills

- Rewards for goal accomplishment

- Better communication skills

- Increase in delegation of responsibilities

Each one of the above team leaders responses regarding changes was related to one or more of the six key dimensions. This relationship will be discussed in the order of the above listing in the following sections.

The additional training in leadership skills response reflected a need for further knowledge in the field of leadership that crossed all six of the key dimensions. Acknowledgement of this need was rather surprising considering the superior attitude of being the best within the field. However, it was comforting knowing that there was some notion of not necessarily being perfect and that room for improvement existed.

The rewards for goal accomplishment response revealed a need to improve the manage performance dimension. This was consistent with the survey findings in this dimension in which three out of the seven items reflected negative correlation with room for improvement. This dimension relates not only to the team, but also to a need for the greater organization to enhance recognition for achievement.

The responses better communication skills and increase in delegation of responsibilities fall within the collaborative climate dimension. This dimension reflected only three out of the ten survey items as having negative correlation. However, improving communications is an area that can only further improve leadership competencies. Learning how to delegate assignments to competent team members will result in additional time for the team leader to focus on other issues, thereby potentially strengthening both the team leaders and members as well.

The team members responses to the first item regarding strengths included the following ten in order of the one most commonly selected:

- Open communications
- Technical knowledge
- Interpersonal skills
- Safety consciousness
- Team player
- Integrity
- Patience
- Enthusiasm
- Compassion
- Organizational skills

Each one of the above team members responses regarding strengths was related to one or more of the six key dimensions. This relationship will be discussed in the order of the above listing and will be related to the team leader responses discussed above in the following sections.

Open communications are key to the collaborative climate dimension. Seven of the ten survey items with positive correlation confirmed this strength. This strength was also listed as the fourth strength within the team leader listing. From both the team leader and team member responses, it was understandable that a collaborative climate existed.

The technical knowledge response coincides exactly with the demonstrate sufficient technical know-how dimension. It is puzzling and inconsistent that the team members would identify technical knowledge as a strength when three of the five items in the survey reflected negative correlation indicating an inconsistent perception between team leaders and team members. This result may be biased due to the small number of team leaders in the sample.

The interpersonal skills response is related to the focus on the goal and ensure a collaborative climate dimensions. Both of these dimensions reflected significant positive correlation in the survey with the focus on the goal dimension resulting in five out of six items and the collaborative climate dimension resulting in seven out of ten items. Additionally the interpersonal skills response was the number one strength on the team leaders lists, thereby substantiating these skills as an important strength.

The safety consciousness response is related to the build confidence dimension. This dimension revealed a positive correlation with six out of seven items supporting the perception that confidence was apparent within the teams. The fact that safety consciousness was even listed as the number four perceived strength amongst the team members within a dangerous industry is very comforting.

The next six-team member identified strengths (team player, integrity, patience, enthusiasm, compassion, and organizational skills) are all related to the build confidence dimension. This dimension reflected a very strong correlation with six of the seven being positive within the survey. Additionally integrity and organizational skills were identified as the fifth and seventh responses on the team leaders listings.

The team members responses to the second item regarding changes included the following in order of the one most commonly selected:

- Improved communications
- Rewarding team accomplishments

- Establishment of performance standards

- Less favoritism within the team

- Providing direction

- Collaborative decision making

Each one of the above team members responses regarding changes was related to one or more of the six key dimensions. This relationship will be discussed in the order of the above listing and will be related to the team leader responses discussed above in the following sections.

Both the number one and number two listed changes, improved communications and rewarding team accomplishments, were also identified as important changes, number three and two, respectively, within the team leaders listings. This concurrence further supports the perceptions between the team leaders and team members that these changes are necessary to improve the effectiveness of the team leader.

The improved communication response falls within the collaborative climate dimension. This dimension reflected only three out of the ten survey items as having negative correlation. However, the need to improve communications is an area that can only enhance the rapport between the team leader and team members in the pursuit of a collaborative climate.

The rewarding team accomplishment, establishment of performance standards, less favoritism within the team, and providing direction responses revealed a need to improve the manage performance dimension. This was consistent with the survey findings in this dimension in which three out of the seven items reflected negative correlation with room for improvement. This dimension reflected important needs that should be corrected in order inspire other teams and individuals to pursue the expected standard and defined goals, and further recognize top-notch performance fairly.

The collaborative decision making response falls within the collaborative climate dimension. This dimension reflected only three out of the ten survey items as having negative correlation. However, the need to improve collaboration is a goal that can only improve the relationships between the team leader and team members in furthering the collaborative climate, while improving productivity and team effectiveness.

Managers Performance Rating Survey Analysis

Four managers of the team leaders provided a set of effectiveness ratings for each of the teams surveyed. Each manager was asked to evaluate the effectiveness of the team by rating each team on a 4-point scale. The highest rating was a 4.0. The managers averaged

performance ratings are summarized in Table 14.

Table 14 Managers team effectiveness ratings

Team	Overall Rating
1	3.1
2	2.75
3	4.0
4	3.23
5	3.25
6	3.95
7	3.252
8	2.751
9	2.15
10	2.5
11	3.52
12	3.65
13	3.5

Each of the managers providing a team rating was asked to identify

the three most important criteria for an effective team. The responses

included the following in order of frequency selected:

- Teamwork

- Strong leadership

- Sound communications

- Strong team foreman to inspire the team

- Mature team leader

- Mutual respect among team personnel

- Reliability and competence of team members

Comparison of Managers Performance Rating to Production Numbers

Actual production levels of the teams being surveyed for the same

period as the survey timeframe resulted in the following listing based on

production performance:

Table 15 Managers rankings versus production rankings

Team	Managers Rankings	Production Ranking
1	9	2
2	11	11
3	1	1
4	8	10
5	7	4
6	2	7
7	6	3
8	10	6
9	13	13
10	12	12
11	4	5
12	3	9
13	5	8

Four out of the thirteen teams or 31% of the teams ranked were consistent

between the rankings based on the managers performance rankings as

compared with the production rankings. This statistic clearly shows that

the assessment of effectiveness cannot be based solely on production.

There are many other factors and variables that come into play that make

this analysis much more complicated than a simple comparison.

Chapter 5

Summary, Conclusions, and Recommendations

This chapter discusses and summarizes the study and the results presented in Chapter 4. The initial discussion focuses on the relationship between team effectiveness, manager performance ratings, and production. The hypotheses developed in Chapter 1 are addressed next. A discussion of the limitations of the research and findings as well as discussion of future research concludes this chapter and the paper.

Manager Ratings, Production, and Team Effectiveness

Manager performance ratings were obtained for all teams involved in this study. The managers surveys were a subjective performance rating of a team s effectiveness. This rating was compared to a rating derived from an objective measure of production and no significant correlation was found between the two.

Certainly one of the reasons for the differences between the two ratings may be due to the comparison of a subjective survey instrument to an objective measure of production. The manager s performance rating was completely subjective in that it asked for a ranking of teams through the use of a 4-point scale. Possibly a scale of 1.0 to 4.0 was too limiting and did not afford enough latitude to allow the managers to be more discriminatory in their grade assignment. The managers were not restricted in limiting the number of teams that were assigned a 4.0 nor were they forced to distribute the teams in their charge across the entire range of the scale. Additionally the managers rankings may have involved more than just production and consequently could have been influenced from many different perspectives. Lastly, subjective rankings are prone to interpretation differences. In this case, managers may not agree as to what constitutes effectiveness because their definitions differ.

The ranking on the production instrument was based solely on actual production numbers, and the relative ranking of the team was based on that number as compared to the other teams. As addressed in the limitations section of Chapter 1, many uncontrollable variables come into play in the performance of a UXO tasking. Therefore, even though the measurement of the production instrument was objective because it focused on the completion and accounting of a unit of work, it may have been biased by variables.

Hypotheses and Summary of Findings

In the formulation of this study, the following hypotheses were based on Frank LaFasto and Carl Larson s (2001) six key dimensions. Through a comparison of survey response results between team leaders and team members, the following hypotheses were tested through the use of two survey questionnaires developed by Frank LaFasto and Carl Larson (2001, pp. 151-154):

1. Clearly defining the team goal contributes to team effectiveness.

2. A collaborative climate increases productivity and team effectiveness.

3. Confidence of team members enhances performance, which results in increased productivity and effectiveness.

4. Job knowledge contributes to improved performance.

5. Establishment of priorities helps achievement of goals.

6. Clear and recognized performance objectives enhance focus and result in better performance.

Given the large number of relationships that were tested in this analysis, the significant findings have been summarized in Table 16 by the research hypothesis and each survey item. Positive and significant relationship is depicted as a + ; no relationship as a 0. A detailed analysis of the findings for each hypothesis within the specific dimension follows in the subsequent sections.

Table 16 Summary of research hypotheses and findings

Hypotheses	Survey Item Number	Findings
Clearly defining the team goal contributes to team effectiveness.	1	+
	2	+
	3	+
	4	+
	5	+
	6	0
A collaborative climate increases productivity and team effectiveness.	7	+
	8	+
	9	+
	10	0
	11	+
	12	0
	13	+
	14	0
	15	+
	16	+
Confidence of team members enhances performance, which results in increased productivity and effectiveness.	17	0
	18	+
	19	+
	20	+
	21	+
	22	+
	23	+
Job knowledge contributes to improved performance.	24	+
	25	0
	26	0
	27	0
	28	+
Establishment of priorities helps achievement of goals.	29	+
	30	+
	31	+
	32	0
	33	+

Hypotheses	Survey Item Number	Findings
Clear and recognized performance objectives enhance focus and result in better performance.	34	+
	35	+
	36	+
	37	0
	38	+
	39	0
	40	0

Analysis of the Dimension: Focus on the Goal

The first hypothesis, clearly defining the team goal contributes to team effectiveness, explored the relationship between the team leaders and the team members perspectives as to whether there was focus on the team goal. Five out of the six survey items reflected positive correlation and agreement between the team leaders and team members that there existed focus on the goal. However, one of the items regarding whether or not the team leaders provided adequate explanation as to adjustments to the team goal revealed a negative correlation of -0.2730 and indicated that there was room for improvement in this area. With means of 3.6875 and 3.7137 for LG6 and G6 respectively, out of a score of 4.0, the disagreement was not considered significant.

There are many reasons why a team may experience changes in its goal. Some of the reasons are internal. Problematic team interpersonal relationships may create loss of harmony resulting in a diffused focus. Replacement of personnel could dilute team capability. A change in the goal may be the best approach in optimizing the team s work effort and maximizing production. Sometimes the pressures are external and may be due to competitive influences or new direction from the client. In any case, the team leaders in this study understood the importance for the team leader being the leader of the team. Additionally the leaders recognized that as the change agents of the teams, it was their responsibility to communicate change effectively so that the entire team was cognizant of the reasons for the change. Without this communication, cohesion within a team would diminish and collaboration would falter.

The importance of being focused on the goal cannot be over emphasized. It is focus that provided the team's objective or purpose in an effort. The team leaders provided a clear vision of the team's goal so that appropriate planning, organizing, and staffing could follow. Then the team leaders clearly communicated the plan to the team so that all team members were aware of the objectives and the approaches to the successful attainment of the goal. It is during communications such as these that a team leader provided the confidence required to enlist the support and commitment of team personnel towards a common goal. Reinforcement of the team's goal was necessary, especially after periods of disillusion, and occurred as frequently as necessary to ensure compliance. A fresh approach to conveying the message assisted in leaving an impression and enabling better retention. A word of caution is that unclear direction to team members may result in confusion, chaos, loss of man-hours, loss of resources, and ultimately inefficiencies and ineffectiveness resulting in schedule delays and loss of production.

Maintaining focus on the common team goal was essential to maximizing production. Acknowledging distracters that have the potential of diffusing focus was important. Confronting team members preoccupied with individual goals that conflicted with the team goal occurred in a swift and conclusive manner. While berating an individual publicly is not acceptable, expressing displeasure of wayward conduct to the team as a unit is recommended, so that peer pressure serves as a deterrent to compromising the team focus. Political motivations and influences were set aside because goal attainment was the objective. Involvement of the manager as a shield to protect the team from external factors was used as necessary to reduce external influences.

Not only are the team leaders responsible for the conduct of the team members, but also they are responsible for assisting the team members in their responsibilities and execution of their duties and obligations toward the team. As the conductor of the team and coach of the personnel, the team leaders provided guidance and compassion toward team members so that their focus was aligned with the team's objective.

Analysis of the Dimension: Ensure a Collaborative Climate

The second hypothesis, a collaborative climate increases productivity and team effectiveness, explored the relationship between the team leaders and the team members perspectives as to whether a collaborative team climate existed. Seven out of the ten survey items reflected positive correlation and agreement between the team leaders and team members that a collaborative climate existed. However, three of the ten items revealed disagreement between the team leaders and the team members.

Item number 10 revealed disagreement regarding the existence of chronic problems within the team as perceived by the team members. Correlation between LCC10 and CC10 was negative at -0.0638 with a significance probability of 0.8143; there was room for improvement in this area. With means of 3.625 and 3.6669 for LCC10 and CC10 respectively, out of a score of 4.0, the disagreement was not considered significant.

The Cronbach alpha for this item for the team leader was barely acceptable when rounded to the closest tenth to 0.7 from 0.6678; for the team member it was very strong at 0.998. However, the term "chronic problems" may have caused some differences in interpretation as to what constituted a "chronic" problem. The likelihood that a team leader would be more sensitive to this confusion was greater based on background experience as compared to that of a team member. Additionally the admittance of a problem that could not be controlled by the team leader who was in charge of the team and responsible for resolving issues was less than that perceived by a team member.

Item number 12 revealed disagreement regarding the recognition and rewarding of behavior that inspired a supportive and open team climate as perceived by the team members. Correlation between LCC12 and CC12 was negative at -0.1692 with a significance probability of 0.5311; there was room for improvement in this area. With means of 3.8125 and 3.7525 for LCC12 and CC12 respectively, out of a score of 4.0, the disagreement was not considered significant.

The importance of recognizing collaborative behavior cannot be over-emphasized. This type of recognition was conducted publicly so that other team members could join in the recognition of their peers. Any form of acknowledgement served as an uplifting tool for the person being recognized and a motivational tool for the others observing the celebration. Because it is human nature to desire strokes, the more recognition given the more people wanted. However, a word of caution was conveyed to team leaders to ensure that the recognition was consistent and appropriately given. Nothing demoralized a crew more as when recognition was meaningless.

Item number 14 revealed disagreement regarding the allowance of structure, systems, and processes to interfere with the achievement of teams goals as perceived by the team members. Correlation between LCC14 and CC14 was negative at -0.1458 with a significance probability of 0.5902; there was room for improvement in this area. With means of 3.4375 and 3.4494 for LCC14 and CC14 respectively, out of a score of 4.0, the disagreement was not considered significant.

Outside influences can be very disruptive to a team. It was extremely important that the team leader stay tuned to potential forces that would undermine the team effort. The team leader needed to set the example for others to emulate. Openness and honesty was a must. There were very few issues that the team leaders were unwilling to discuss openly with the team members. This was important because trust will be compromised and commitment will wane if the team senses that the team leaders were not being frank.

Instilling a safe climate that enabled team members to speak out and present their views regarding team issues without fear of reprisal was one of the most important duties performed by the team leaders. The team leaders fostered an open and safe climate by encouraging the contribution of team member's thoughts during positive exchanges between members. It was the team leader's responsibility to eliminate perceptions and actions that inhibited open communication between members. Unacceptable and non-collaborative behavior was immediately dealt with and was not allowed to continue. A collaborative climate enabled team members to provide input to resolve problems that were impeding goal attainment. This involvement encouraged freedom of ideas and better understanding of issues. Oftentimes team members brought input that was founded on close observation and had already been through some discussion with other team members.

A team leader needed to be cautious of silence among members. Sometimes silence was an indicator of internal team problems that needed to be addressed. If the team member was unable to determine the cause of a festering problem, other managers were consulted for suggestions and assistance.

Problem solving utilizing a team collaborative approach is best. Involvement of team personnel in the decision making process built esprit de corps and a sense of contribution that increased sharing of problems and acceptance of unified solutions.

The need for controlling the teams required balance and understanding. Team leaders had to set aside egos that inhibited open communication and collaboration. While the ultimate control of the team rested with the team leaders, the art of listening, analyzing, and executing actions that represented collaborative decisions reflected a willingness to evaluate suggestions with an open mind and trust. This approach resulted in a healthy, safe, and collaborative environment that benefited the entire team.

Analysis of the Dimension: Build Confidence

The third hypothesis, confidence of team members enhances performance which results in increased productivity and effectiveness, explored the relationship between the team leaders and the team members perspectives as to whether a confidence building environment existed. Six out of the seven survey items reflected positive correlation and agreement between the team leaders and team members that a confidence-building environment existed.

Item number 17 revealed disagreement regarding the focus of the team leader on achieving results within the team as perceived by the team members. Correlation between LCC17 and CC17 was negative at

-0.2245 with a significance probability of 0.4031; there was room for improvement in this area. With means of 3.875 and 3.9331 for LCC17 and CC17 respectively, out of a score of 4.0, the disagreement was not considered significant.

The team leaders were the strongest proponents on the team that were focused on achieving success by setting the example for the others to emulate. The tone of each of the teams was dependent on the positive influence exerted by the team leaders. The team leaders needed to be optimistic in approach to issues and take advantage of opportunities to excel when presented. The teams successes in meeting goals and objectives were viewed as each individual member's success; even failures were shared. However, it was the successes that built the confidence levels of the team members to continue to drive toward more successes.

A collaborative team leader provides the inspiration required by accomplishing almost the impossible by encouraging the support of the team members. This support was obtained through respect of the team members and trust in their abilities. It was the positive reinforcement of actions that propelled people to perform to the best of their ability. Trust in team members was exhibited by assigning meaningful tasks concomitant with individual abilities for accomplishment. The prudent use of talents was capitalized whenever possible.

Clarity in the objective was important to focus energies towards goal accomplishment. Issues that had a direct or indirect bearing on a team member would be openly discussed so that misperceptions did not occur. Collaborative solutions to issues were sought whenever possible to enhance the feeling of contribution and worth to the team.

Fair and impartial treatment of team members is a must. Any impartial actions would quickly be noticed and served to destroy any and all positive attributes. All team members needed to feel safe, wanted, and appreciated for their contributions to the teams. Opportunities to recognize top performance were pursued and capitalized. Accolades were publicly made so that other members of the team could extend congratulations.

Analysis of the Dimension: Demonstrate Sufficient Technical Know-how

The fourth hypothesis, job knowledge contributes to improved performance, explored the relationship between the team leaders and the team members perspectives as to whether sufficient technical know-how existed. Only two out of the five survey items reflected positive correlation and agreement between the team leaders and team members that job knowledge contributed to improved performance.

Item number 25 revealed disagreement regarding the experience level of the team leaders in the technical aspects within the team's goal as perceived by the team members. Correlation between LT25 and T25 was negative at -0.1698 with a significance probability of 0.5296; there was room for improvement in this area. With means of 3.9375 and 3.8431 for LT25 and T25 respectively, out of a score of 4.0, the disagreement was not considered significant.

Team leaders needed to demonstrate strong leadership skills that support the achievement of the teams goals. These provided a sense of security to the teams in knowing that the team leaders were seasoned individuals that were able to provide sound judgement and prudent approaches towards mission accomplishment. This is especially true in hazardous professions, such as the UXO industry, in which the team leaders experiences in the field can make the difference between success and failure or even survival or death.

Item number 26 revealed disagreement regarding the team leaders openness to accept advice from team members who may be more knowledgeable in the technical aspects of achieving the team's goal as perceived by the team members. Correlation between LT26 and T26 was negative at -0.0793 with a significance probability of 0.7704; there was room for improvement in this area. With means of 3.9375 and 3.7644 for LT26 and T26 respectively, out of a score of 4.0, the disagreement was not considered significant.

In technical fields, particularly those in which a single mistake can result in serious injury or death such as that found in the UXO industry, it is natural for a team leader to be reluctant to accept advice from a non-UXO qualified person. That is not to say that one should not be receptive to advice nor portray a negative attitude towards the acceptance of suggestions, but certainly the advice needs to be evaluated and if it appears to be valuable used. It was important for all members of the team to recognize that the ultimate decision rested with the team leaders as the accountability and responsibility belonged solely to the team leaders. Additionally team members frequently had suggestions that merited open discussion, so that a climate of collaboration was maintained even though the final authority rested with the team leader.

Item number 27 revealed disagreement regarding the capability of the team leaders to analyze complex issues related to the team's goal as perceived by the team members. Correlation between LT27 and T27 was negative at -0.3062 with a significance probability of 0.2487; there was room for improvement in this area. With means of 3.875 and 3.8688 for

LT27 and T27 respectively, out of a score of 4.0, the disagreement was not considered significant.

This issue is one of trust in the team leaders abilities to deal with the complex issues. Team leaders needed to demonstrate proficiency in the technical aspects of the job. If knowledge was lacking, then it was important that the team leaders pursued advice from others that could assist with an issue. In no case should team leaders be timid to admit the lack of knowledge on an issue or merely just ignore an issue, because it will not just go away. It is not unusual that even among the most effective team leaders that areas exist in which their knowledge was less than satisfactory. Furthermore, team members expected the team leader to deal with and resolve issues so that mission accomplishment was achieved.

Conduct of UXO clearance operations involved many technical issues that are not just common sense. Unless one is schooled in ordnance recognition, hazards, identification, and disposal, operational decisions needed to be deferred to those that have proven competence in those areas through completion of formal training. All of the team leaders are UXO qualified personnel who were graduates from formal Explosive Ordnance Disposal training that covered all areas encountered during clearance operations. Therefore, it stands to reason that each team leader understood the technical issues associated in achieving team goals. Competencies among the team leaders varied, but that was to be expected. However, a minimum level of competency was ensured through the successful completion of contractor training which included a refresher on UXO. A minimum score needed to be attained before an individual was even considered for selection to the UXO team leader level.

An ongoing process of surveillances and inspections by both the contractor quality control and client quality assurance personnel validated credibility of team leaders by others who may be knowledgeable. Additionally the area managers and safety personnel were in constant contact with team personnel and maintained a close eye on the effectiveness, efficiency, and safe conduct of operations. The level of oversight on this specific project was substantial. However, oversight is not an unusual occurrence on UXO clearance projects because of the varied stakeholders and interested agencies monitoring progress and ensuring compliance with various regulations.

Analysis of the Dimension: Set Priorities

The fifth hypothesis, establishment of priorities helps achievement of goals, explored the relationship between the team leaders and the team members perspectives as to whether the setting of priorities was considered in the pursuit of the team goal. Five out of the six survey items reflected positive correlation and agreement between the team leaders and team members that priorities were considered. However, one of the items regarding whether the team leaders diluted the team's efforts with too many priorities revealed a negative correlation of -0.0477 with a significance probability of 0.8606 and indicated that there was room for improvement in this area. With means of 3.6875 and 3.6769 for LP32 and P32 respectively, out of a score of 4.0, the disagreement was not considered significant.

The perception that the team was inundated with too many priorities was a common complaint that usually was based on erroneous information or miscommunication. Again the importance of clarity of communications between the team leaders and the team members cannot be over-emphasized. Diluting the team's focus with too many priorities resulted in reduced effectiveness and confusion. Team leaders needed to be vigilant to note conflicts in tasking that resulted in competing pressures to accomplish more than what could be accomplished with the resources and manpower assigned.

Proper planning with establishment of priorities and schedules assisted in limiting distractions that could reduce focus. Priorities needed to be manageable considering the assets assigned and tasking involved.

Schedules needed to be realistic and preferably based on historical precedents for similar efforts. Adjustments to the schedule were made as necessary in order to manage progress, reallocate assets, and predict task completion.

Maintaining a collaborative approach to assigning top priorities enabled involvement of personnel at all levels. Interface during priority discussion sessions enabled a team leader to recognize abilities of team members through communication exchanges and identify those that needed some assistance. Additionally, those sessions enabled a team leader to clearly define the priority and ensured that each team member's understanding of the priority was the same.

It was the team leaders responsibility to communicate and maintain a focus on priorities. Team leaders that were wishy-washy and did not actively ensure focus on the priority did not meet deadlines, lost control of the team, and jeopardized achievement of the goal. Team members expected leaders that were devoted and dedicated in accomplishment of goals.

Unfortunately priorities did change for various reasons. Whatever the reasons, the team leaders needed to clearly communicate the explanation to the team members as soon as possible. Complete understanding of the reasons for the change was ascertained from each team member to avoid future misperceptions. Even if the change was a minor adjustment to the priority, this change needed to be described and understood. Impact to each team member's approach to task accomplishment needed to be evaluated and adjustments made to ensure that efficiency was not compromised and achievement was still possible.

Analysis of the Dimension: Manage Performance

The sixth hypothesis, clear and recognized performance objectives enhance focus and result in better performance, explored the relationship between the team leaders and the team members perspectives as to whether clear and recognized performance objectives existed. Four out of the seven survey items reflected positive correlation and agreement between the team leaders and team members that clear performance objectives existed. However, three of the seven items revealed disagreement between the team leaders and the team members.

Item number 37 revealed disagreement regarding the assessment of the team member's collaborative skills and the results achieved by the team leaders as perceived by the team members. Correlation between LMP37 and MP37 was negative at -0.0406 with a significance probability of 0.8814; there was room for improvement in this area. With means of

3.6875 and 3.7163 for LMP37 and MP37 respectively, out of a score of 4.0, the disagreement was not considered significant.

Being able to conduct an assessment of collaborative skills that team members bring to a team require a team leader to be knowledgeable in the area of the tasking. UXO clearance is not rocket science. All UXO qualified team leaders possessed the requisite knowledge to assess performance. Oftentimes it was not that the assessment was not being accomplished, but rather the results of the assessment were not always communicated to the team members. It was essential that constructive feedback be provided to fine tune both individual and team performance and procedures. It was the team leaders responsibility to show interest, stroke team members to keep them motivated, and maintain a collaborative environment that thrived on open communication. Acknowledgement of results was publicly recognized because it was that type of attention that inspired others to perform.

Item number 39 revealed disagreement regarding the willingness of the team leaders to confront and resolve issues associated with inadequate performance by team members as perceived by the team members. Correlation between LMP39 and MP39 was negative at -0.2462 with a significance probability of 0.3580; there was room for improvement in this area. With means of 3.8125 and 3.6275 for LMP39 and MP39 respectively, out of a score of 4.0, the disagreement was not considered significant.

Confronting and resolving performance issues was an area that caused tremendous turmoil within a team. This responsibility rested with the team leaders to ensure that all team members were pulling their weight and that no one was allowed to skate. If a single allowance were made, an example that would plague the effectiveness of the team would be set. Team members would then use this case at will to negotiate and bargain with the team leaders. After all, team members were paid hourly to perform in a satisfactory manner. Fairness was maintained so the team leader would not be in a weak position to enforce appropriate actions. Standards of conduct were essential to good order and discipline. Enforcement of standards was important, as otherwise the establishment of a standard was nonsensical. Non-performers were counseled and put on notice so that others did not fall into the same rut thereby jeopardizing goal accomplishment.

Item number 40 revealed disagreement regarding the team leaders recognizing and rewarding superior performance as perceived by the team members. Correlation between LMP40 and MP40 was negative at -0.1950 with a significance probability of 0.4693; there was room for improvement in this area. With means of 3.9375 and 3.625 for LMP40 and MP40 respectively, out of a score of 4.0, the disagreement was not considered significant.

Public recognition of superior results was important not only to the individual that had achieved the accomplishment, but also to the rest of the team members to keep them motivated and inspired to perform to the best of their abilities. It was of utmost importance to ensure that the recognition was provided for valid and meaningful superior results consistent with the teams goals, as otherwise the recognition became more of a mockery than a meaningful event. Fairness in the administering and level of recognition was maintained for recipients. It was the team leaders duties to take the time and effort to provide recognition to those personnel that deserved it. An incentive program with an active recognition program inspired personnel to perform.

Establishment of performance standards oftentimes was dictated from a level higher than the team level, but in those instances in which it was not, it was up to the team leader to create the standard. Nothing precludes the team leader from establishing additional standards for the team as long as they are reasonable. Enforcement of standards was without item the sole responsibility of the team leader. Additionally the team leader ensured that the expected standard was clearly understood by the team members. Any items needed to be thoroughly addressed so no doubt remained as to the expectations. And of course the expectations needed to be consistent with the standards.

Involvement of the team in establishing a set of values that guided performance was consistent with collaborative behavior and was encouraged. The team leaders as the leaders of the teams set the example and "walked the talk" for others to emulate. Besides taking in consideration team member ideas and opinions, involvement of the team members provided an added benefit by making the team members feel as part of the decision making process, which may have resulted in upholding the values to a greater extent. Additionally, peer pressure would force violators to think twice before deviating from the accepted values. The team leaders used these team agreed upon values as the standards to hold team members to when problems arose.

Constructive feedback was essential to good performance. Team members needed to know that their performance was being scrutinized, held to a standard, and evaluated. Periodic evaluations were necessary to validate that performance was meeting expectations and to identify areas for improvement that would enable the individual and the team to achieve their goal. Goals and objectives needed to be reiterated during periodic reviews and performance evaluation sessions. A plan to resolve weak areas with periodic due dates for accomplishing certain events needed to be defined. Personal development options and future goals were discussed. It was not uncommon for teams that perceived that they received more team level feedback and recognition to have higher levels of satisfaction in their efforts towards goal accomplishment.

It takes a tremendous amount of effort and energy to become an effective leader. Leadership is an opportunity and privilege that only the dedicated and very best should have. Those that are not up to the challenge should not attempt to execute the duties and responsibilities of a leader.

Implications of Findings

This research paper adopted the six key dimensions (focus on the goal, ensure a collaborative climate, build confidence, demonstrate sufficient technical know-how, set priorities, and manage performance) presented by Frank LaFasto and Carl Larson (2001) as essential elements to team leader effectiveness. The findings further substantiated these dimensions as necessary elements to achieve effectiveness and productivity and eventually success in the UXO industry through collaboration between the team leaders and the teams. This research has provided information that will be beneficial to team leaders in the performance of their duties and responsibilities as leaders of UXO clearance teams. Adoption and use of Frank LaFasto and Carl Larson s (2001) six key dimensions will provide an outstanding guide for team leaders to use when leading their teams. The findings identified in this study provide some empirical support and insight into the strengths and weaknesses found in one case study. The dimension that was found to be the weakest, the demonstration of sufficient technical know-how, was actually expected to be one of the strongest. Overall, one can conclude that this research has made an important contribution towards defining: how a leader creates and handles an effective unexploded ordnance team.

Additionally this study should provide some insight to management as to the importance of selecting effective UXO clearance team leaders. A review of the selection process should encompass more than just a focus on the number of years of UXO experience. Prior experience in leadership roles should be considered and weighed heavily in the selection. References should be contacted to confirm performance at previous locations. Job interviews should provide hypothetical situations and evaluate decisions from prospective candidates. Selection decisions should not be based solely on resumes.

Future Research

There are a number of interesting research and methodological issues that merit further future research. Additional studies of UXO teams involved in other UXO clearance projects needs to be undertaken for comparison purposes. Since this study is the first of its kind in the UXO industry, there were no other studies to use for comparison. While this study established the precedent for others to follow, it is anticipated that additional studies may be able to focus on the weak dimension areas to further substantiate those findings.

Manager performance ratings examination of the team leaders should be explored. Since the focus of this research was one level lower, the team leader to the team member, that area was not evaluated.

Expanding the evaluation of the team leader with a larger statistical sample would be interesting as well. Focus on prior leadership experience and the influence of that experience would provide another interesting aspect. A comparison of those results to the findings of this research would further substantiate the outcomes.

Further studies that include more explicit measures to characterize the various tasks being performed as they influence leadership techniques within the teams would provide yet another relationship dimension that was not addressed in this study.

In summary this study only scratches the surface in the body of knowledge that can be obtained with further research of the UXO industry. With the continuous growth in this industry due to UXO contamination by warring nations or military training, it seems important that additional studies are conducted to explain procedures and techniques used to further refine processes that will result in increased efficiency and effectiveness of clearance. As costs continue to rise, better management and leadership skills are required and expected. Unfortunately there is no cookbook that adequately explains approaches to every incident and situation that may occur. The foundation of this business has been established through blood, sweat, and tears. Consequently this industry is ripe for further studies that will solidify methodologies and technologies and minimize equipment failures and personnel losses.

References

Amason, A. C., & Thompson, K. R. (1995). Conflict: An important dimension in successful management teams. *Organizational Dynamics, 24*(2), 20-35.

Antonioni, D. (1994, May). Managerial roles for effective team leadership. *Supervisory Management, 39*(5), 3.

Avery, C. M. (1999, April). All power to you: Collaborative leadership works. *Journal for Quality and Participation, 22*(2), 36-50.

Barnett, D., & McKowen, C. (1998, April). A school without a principal. *Educational Leadership, 55*(7), 48-49.

Bass, B. M., & Stogdill, R. M. (1990). *Bass and Stogdill s handbook of leadership*. New York: Free Press.

Bennis, W. G., & Biederman, P. W. (1997). *Organizing Genius: The secrets of creative collaboration*. Reading, Massachusetts: Addison-Wesley.

Bennis, W. G., & Nanus, B. (1985). *Leaders: the strategies for taking charge*. New York: Harper & Row.

Benson-Armer, R., & Stickel, D. (2000). Successful team leadership is built on trust. *Ivey Business Journal, 64*(5), 20-25.

Bethel, S. M. (1990). *Making a difference: Twelve qualities that make you a leader*. New York: Putnam.

Blake, R. R., & Mouton, J. S. (1964). *The managerial grid*. Houston, Texas: Gulf.

Bowers, D. G., & Seashore, S. E. (1967). Peer leadership within work groups. *Personnel Administration, 30*, 45-50.

Bradford, D. L., & Cohen, A. R. (1984). *Managing for excellence: The guide to developing high performance in contemporary organizations*. New York: Wiley.

Bragg, T. (2000, February). How to build great work teams needs full planning. *Hudson Valley Business Journal, 11*(1), 26-27.

Brown, T.M., & Miller, C.E. (2000, April). Communication networks in task-performing groups. *Small Group Research, 31*(2), 131-157.

Bryman, A. (1986). *Leadership and organizations*. Boston: Routledge & Kegan Paul.

Burke, P. J. (1966). Authority relations and disruptive behavior in small discussion groups. *Sociometry, 29*, 237-250.

Careless, S. A., & de Paola, C. (2000, February). The measurement of cohesion in work teams. *Small Group Research, 31*(1), 71-87.

Cattell, R. B. (1957). A mathematical model for the leadership role and other personality-role relations. In M. Sherif & M. O. Wilson (Eds.),

Emerging problems in social psychology. Norman: University of Oklahoma.

Cohen, S. G., & Bailey, D. E. (1997). What makes team work: Group effectiveness research from the shop floor to the executive suite. *Journal of Management, 23*(3), 239-291.

Componation, P.J., Utley, D.R., & Swain, J.J. (2001, December). Using risk reduction to measure performance. *Engineering Management Journal, 13*(4), 27-34.

Crosby, P. B. (1986). *Running things: The art of making things happen*. New York: McGraw-Hill.

Csoka, L. S. (1998). *Bridging the leadership gap*. New York: The Conference Board.

Dessler, G. (2001). *Management: Leading people and organizations in the 21st century*. Upper Saddle River, New Jersey: Prentice Hall.

Devine, D. J., & Clayton, L. D. (1999, Dec). Teams in Organization. *Small Group Research, 30*(6), 34-38.

Dimock, H. G. (1970a). *Leadership and group development*. (Part I Factors in working with groups). Montreal, Quebec: Sir George Williams University.

Dimock, H. G. (1970b). *Leadership and group development.* (Part II How to observe your group). Montreal, Quebec: Sir George Williams University.

Fiedler, F. E. (1958). *Leader attitudes and group effectiveness; final report of ONR project NR170-106, N6-ori-07135.* Urbana, Illinois: University of Illinois Press.

Fitz-Enz, J. (1997, August). Measuring team effectiveness. *HR Focus, 74*(8), 3-5.

Gilpin, A. (2002, March 21). The power of soulpower. *Leisure and Hospitality Business*, 12.

Gist, M. E., Locke, E. A., & Taylor, S. M. (1987). Organizational behavior: Group structure, process, and effectiveness. *Journal of Management, 13*(2), 237-257.

Goodman, P. S., & Leyden, D. P. (1991, August). Familiarity and group productivity. *Journal of Applied Psychology, 76*(4), 578-586.

Guzzo, R. A., & Dickson, M. W. (1996). Teams in organizations: Recent research on performance and effectiveness. *Annual Review of Psychology, 47*(1), 307-338.

Harrington-Mackin, D. (1994). *The team building tool kit: Tips, tactics, and rules for effective workplace teams.* New York: American Management Association.

Hackman, J. R., & Walton, R. E. (1986). Leading groups in organizations. In P. S. Goodman & Associates (Eds.), *Designing effective work groups* (pp. 72-119). San Francisco: Jossey-Bass.

Hays, S. (1999, December). Our future requires collaborative leadership. *Workforce, 78*(12), 30-32.

Heenan, D. A., & Bennis, W. (1999). *Co-leaders: The power of great partnerships.* New York: Wiley.

Heim, D. (1996, June 19). Phil Jackson, Seeker in sneakers. *Christian Century, 113*(20), 654-656.

Henderson, D., & Green, F. (1997). Measuring self-managed workteams. *Journal for Quality & Participation, 20*(1), 52-57.

Hersey, P., Blanchard, K. W., & Johnson, D. E. (2001). *Management of organizational behavior: Leading human resources.* Upper Saddle River, New Jersey: Prentice Hall.

Hollander, E. P. (1978). *Leadership dynamics: A practical guide to effective relationships.* New York: Free Press.

Huey, J., & Colvin, G. (1999, January 11). Staying smart: The Jack and Herb show. *Fortune, 139*(1), 163-165.

Hughes, R. L., Ginnett, R. C., & Curphy, G. J. (2002). *Leadership: Enhancing the lessons of experience.* New York: McGraw-Hill.

Hyatt, D. E., & Ruddy, T. M. (1997). An examination of the relationship between work group characteristics and performance: Once more into the breech. *Personnel Psychology, 50*(3), 553-585.

Ilgen, D. R., & Klein, H. J. (1988). Individual motivation and performance: Cognitive influences on effort and choice. In J. P. Campbell & R. C. Campbell (Eds.), *Productivity in organizations* (pp. 143-176). San Francisco: Jossey-Bass.

Ivancevich, J. M., & Matteson, M. T. (2002). *Organizational behavior and management.* New York: McGraw-Hill.

Janz, B. D., & Colquitt, J. A. (1997). Knowledge worker team effectiveness: The role of autonomy, interdependence, team development, and contextual support variables. *Personnel Psychology, 50*(4), 877-904.

Johnson, R. (1996, April). Effective team building. *HR Focus, 73*(4), 18.

Katzenbach, J. R., & Smith, D. K. (1992). *McKinsey Quarterly*, 4, 128-142.

Kelley, R. E. (1998). *How to be a star at work.* New York: Times Books.

Kirkman, B.L. (2000). Why do employees resist teams? Examining the resistance barrier to work team effectiveness. *International Journal of Conflict Management, 11*(1), 74-92.

Kline, T.J.B., & MacLeod, M. (1996). Team effectiveness: Contributors and hindrances. *Human Systems Management, 15*(3), 183-186.

Knowles, H., & Shepherd, M. (1955). *How to develop better leaders.* New York: Association Press.

Kotter, J. P. (1992). What leaders really do? In M. Syrett & C. Hogg (Eds.), *Frontiers of leadership* (pp.16-24). Cambridge, Massachusetts: Blackwell.

Labich, K. (1992). The seven keys to business leadership. In M. Syrett & C. Hogg (Eds.), *Frontiers of leadership* (pp.225-233). Cambridge, Massachusetts: Blackwell.

LaFasto, F. M. J., & Larson, C. E. (2001). *When teams work best: 6,000 team members and leaders tell what it takes to succeed.* Thousand Oaks, California: Sage.

Lampe, D. (1994). Filling a hole. *National Civic Review, 83*(4), 502-504.

Laroche, L. (2001, April). Teaming up. *CMA Management, 75*(2), 22-25.

Larson, C. E., & LaFasto, F. M. J. (1989). *Teamwork: What must go right, what can go wrong.* Newbury Park, California: Sage.

Lord, R. G. (1977, March). Functional leadership behavior: Measurement and relation to social power and leadership perceptions. *Administrative Science Quarterly, 22*(1), 114-133.

Mahoney, T. A. (1988). Productivity defined: The relativity of efficiency, effectiveness, and change. In J. P. Campbell & R. C. Campbell

(Eds.), *Productivity in organizations* (pp.13-39). San Francisco: Jossey-Bass.

McDonald, B. P., & Hutchenson, D. (1998, March 6). A strong group identity is imperative to a successful work team. *Business Press, 10*(45), 15-17.

Moment, D., & Zaleznik, A. (1963). *Role development and interpersonal competence: An experimental study of role performances in problem-solving groups.* Boston: Harvard University, Division of Research, Graduate School of Business Management.

Northouse, P. G. (2001). *Leadership: Theory and practice.* Thousand Oaks, California: Sage.

Ogbonna, E., & Harris, L. C. (2000, August). Leadership style, organizational culture and performance: Empirical evidence from UK companies. *International Journal of Human Resource Management, 11*(4), 766-788.

Paris, C. R., & Salas, E. (2000, August). Teamwork in multi-person systems: A review and analysis. *Ergonomics, 43*(8), 1052-1075.

Pearce, C.L., Gallagher, C.A., & Ensley, M.D. (2002, March). Confidence at the group level of analysis: A longitudinal investigation of the relationship between potency and team effectiveness. *Journal of Occupational & Organizational Psychology, 75*(1), 115-119.

Pearce, C. L., & Sims, Jr. H. P. (2000). Shared leadership: Toward a multi-level theory of leadership. In M. M. Beyerlein, D. A. Johnson, & S. T. Beyerlein (Eds.). *Team Development* (pp.115-139). New York: Elsevier Science.

Pearce, C.L., & Sims, H.P. (2002, June). Vertical versus shared leadership as predictors of the effectiveness of change management teams: An examination of aversive, directive, transactional, transformational, and empowering leader behaviors. *Group Dynamics: Theory, Research, and Practice, 6*(2), 172-197.

Rawlings, D. (2000). Collaborative Leadership Teams: Oxymoron or new paradigm. *Consulting Psychology Journal, 52*(1), 36-48.

Robbins, H., & Finley, M. (1995). *Why teams don t work: What went wrong and how to make it right*. Princeton, New Jersey: Peterson s/Pacesetter Books.

Roby, T. B. (1961). The executive function in small groups. In L. Petrullo & Bass (Eds.), *Leadership and interpersonal behavior* (pp.118-136). New York: Holt, Reinhart, & Winston.

Roche, J. (1994, February). Motivation: Key to productivity. *Landscape Management, 33*(2), 28.

Rocine, V., & Irwin, D. (1994, October). Make team members responsible for team effectiveness. *CMA Magazine, 68*(8), 28.

Rost, J. C. (1991). *Leadership for the twenty-first century*. New York: Praeger.

Ruel, H. J. M. (2000). Reconsidering our team effectiveness models: a call for an integrative paradigm. In M. M. Beyerlein, D. A. Johnson, & S. T. Beyerlein (Eds.). *Team Development* (pp.173-185). New York: Elsevier Science.

Santora, J. C., & Sarros, J. C. (1995, June). Focus on books: Collaborative leadership   how citizens and civic leaders can make a difference. *Business Horizons, 38*(3), 89-90.

Schul, B. D. (1975). *How to be an effective group leader*. Chicago: Nelson-Hall.

Schultz, H., & Yang, D. J. (1997). *Pour your heart into it: How Starbucks built a company one cup at a time*. New York: Hyperion.

Schutz, W. C. (1961). The ego, FIRO theory, and the leader as the completer. In L. Petrullo & Bass (Eds.), *Leadership and interpersonal behavior* (pp.48-65). New York: Holt, Reinhart, & Winston.

Seaman, D. F. (1981). *Working effectively with task-oriented groups*. New York: McGraw-Hill.

Schermerhorn, J. R., Jr. (2002). *Management*. New York: John Wiley & Sons.

Schermerhorn, J. R., Hunt, J. G., & Osborn, R. N. (2000). *Organizational Behavior*. New York: John Wiley & Sons.

Steckler, N., & Fondas, N. (1995). Building team leader effectiveness: A diagnostic tool. *Organizational Dynamics, 23*(3), 20-35.

Stogdill, R. M. (1959). *Individual behavior and group achievement*. New York: Oxford University Press.

Sundstrom, E., De Meuse, K. P., & Futrell, D. (1990, February). Work teams applications and effectiveness. *American Psychologist, 45*(2), 120-133.

Thomas, K. W. (1976). Conflict and conflict management. In M. D. Dunnette (Ed.), *Handbook of industrial and organizational psychology*. Chicago: Rand McNally.

Ulrich, D., Zenger, J., & Smallwood, N. (1999). *Results-based leadership*. Boston: Harvard Business School Press.

Weterings, S. (1998, July). Bridging the leadership gap. *Management, 45*(6), 21.

Wilson, J. L. (2002, January). Leadership development: Working together to enhance collaboration. *Journal of Public Health Management & Practice, 8*(1), 21-26.

Yukl, G. (1989, July). Managerial leadership: A review of theory and research. *Journal of Management, 15*(2), 251-289.

Zaccaro, S. J., Rittman, A. L., & Marks, M. A. (2001). Team Leadership.

Leadership Quarterly, 12(4), 451-483.

APPENDICES

APPENDIX A

The Collaborative Team Leader
(Team Leader Version)

I.   **Focus on the Goal**

True	More True Than False	More False Than True	False		
☐	☐	☐	☐	1.	I clearly define our goal.
☐	☐	☐	☐	2.	I articulate our goal in such a way as to inspire commitment.
☐	☐	☐	☐	3.	I avoid compromising the team's objective with political issues.
☐	☐	☐	☐	4.	I help individual team members align their roles and responsibilities with the team goal.
☐	☐	☐	☐	5.	I reinforce the goal in fresh and exciting ways.
☐	☐	☐	☐	6.	If it's necessary to adjust the team's goal, I make sure the team understands why.

II.   **Ensure a Collaborative Climate**

☐	☐	☐	☐	7.	I create a safe climate for team members to openly and supportively discuss any issue related to the team's success.
☐	☐	☐	☐	8.	I communicate openly and honestly.
☐	☐	☐	☐	9.	There are no issues that I am uncomfortable discussing with the team.
☐	☐	☐	☐	10.	There are no chronic problems within our team that we are unable to resolve.
☐	☐	☐	☐	11.	I do not tolerate a noncollaborative style by team members.
☐	☐	☐	☐	12.	I acknowledge and reward the behaviors that contribute to an open and supportive team climate.
☐	☐	☐	☐	13.	I create a work environment that promotes productive problem solving.
☐	☐	☐	☐	14.	I do not allow organization structure, systems, and processes to interfere with the achievement of our team's goal.
☐	☐	☐	☐	15.	I manage my personal control needs.
☐	☐	☐	☐	16.	I do not allow my ego to get in the way.

III.   **Build Confidence**

True	More True Than False	More False Than True	False	
☐	☐	☐	☐	17. I ensure that our team achieves results.
☐	☐	☐	☐	18. I help strengthen the self-confidence of team members.
☐	☐	☐	☐	19. I make sure that team members are clear about critical issues and important facts.
☐	☐	☐	☐	20. I exhibit trust by giving team members meaningful levels of responsibility.
☐	☐	☐	☐	21. I am fair and impartial toward all team members.
☐	☐	☐	☐	22. I am an optimistic person who focuses on opportunities.
☐	☐	☐	☐	23. I look for and acknowledge contributions by team members.

IV.   **Demonstrate Sufficient Technical Know-How**

True	More True Than False	More False Than True	False	
☐	☐	☐	☐	24. I understand the technical issues we must face in achieving our goals.
☐	☐	☐	☐	25. I have had sufficient experience with the technical aspects of our team's goal.
☐	☐	☐	☐	26. I am open to technical advice from team members who are more knowledgeable than I am.
☐	☐	☐	☐	27. I am capable of helping the team analyze complex issues related to our goal.
☐	☐	☐	☐	28. I am seen as credible and knowledgeable by people outside our team.

V.   **Set Priorities**

True	More True Than False	More False Than True	False	
☐	☐	☐	☐	29. I keep the team focused on a manageable set of priorities that will lead to the accomplishment of our goal.
☐	☐	☐	☐	30. Team members and I agree on the top priorities for achieving our goal.
☐	☐	☐	☐	31. I communicate and reinforce a focus on priorities.
☐	☐	☐	☐	32. I do not dilute the team's effort with too many priorities.
☐	☐	☐	☐	33. If it's necessary to change priorities I make sure the team understands why.

VI.   **Manage Performance**

	More True Than False	*More False Than True*	
True			*False*

☐	☐	☐	☐	34. I make performance expectations clear.
☐	☐	☐	☐	35. I encourage the team to agree upon a set of values that guides our performance.
☐	☐	☐	☐	36. I ensure that rewards and incentives are aligned with achieving our team's goal.
☐	☐	☐	☐	37. I assess the collaborative skills of team members as well as the results they achieve clearly define our goal.
☐	☐	☐	☐	38. I give useful, developmental feedback to team members.
☐	☐	☐	☐	39. I am willing to confront and resolve issues associated with inadequate performance by team members.
☐	☐	☐	☐	40. I recognize and reward superior performance.

41. What are my strengths as team leader?

42. What one or two changes are most likely to improve my effectiveness as team leader?

SOURCE: Frank M. J. LaFasto and Carl E. Larson, *When team works best: 6,000 team members and leaders tell what it takes to succeed*, pp. 151-152. Copyright © 1996 by Frank M. J. LaFasto, Ph.D. and Carl E. Larson, Ph.D. Reprinted by permission of Frank M. J. LaFasto, Ph. D. and Carl E. Larson, Ph.D.

APPENDIX B

The Collaborative Team Leader
(Team Version)

I. **Focus on the Goal**

True	More True Than False	More False Than True	False	
☐	☐	☐	☐	1. Our team leader clearly defines our goal.
☐	☐	☐	☐	2. Our team leader articulates our goal in such a way as to inspire commitment.
☐	☐	☐	☐	3. Our team leader avoids compromising the team's objective with political issues.
☐	☐	☐	☐	4. Our team leader helps individual team members align their roles and responsibilities with the team goal.
☐	☐	☐	☐	5. Our team leader reinforces the goal in fresh and exciting ways.
☐	☐	☐	☐	6. If it's necessary to adjust the team's goal, our team leader makes sure we understand why.

II. **Ensure a Collaborative Climate**

True	More True Than False	More False Than True	False	
☐	☐	☐	☐	7. Our team leader creates a safe climate for team members to openly and supportively discuss any issue related to the team's success.
☐	☐	☐	☐	8. Our team leader communicates openly and honestly.
☐	☐	☐	☐	9. There are no issues that our team leader is uncomfortable discussing with the team.
☐	☐	☐	☐	10. There are no chronic problems within our team that we are unable to resolve.
☐	☐	☐	☐	11. Our team leader does not tolerate a noncollaborative style by team members.
☐	☐	☐	☐	12. Our team leader acknowledges and rewards the behaviors that contribute to an open and supportive team climate.
☐	☐	☐	☐	13. Our team leader creates a work environment that promotes productive problem solving.
☐	☐	☐	☐	14. Our team leader does not allow organization structure, systems, and processes to interfere with the achievement of our team's goal.
☐	☐	☐	☐	15. Our team leader manages his/her personal control needs.
☐	☐	☐	☐	16. Our team leader does not allow his/her ego to get in the way.

III.   Build Confidence

True	More True Than False	More False Than True	False	
☐	☐	☐	☐	17. Our team leader ensures that our team achieves results.
☐	☐	☐	☐	18. Our team leader helps strengthen the self-confidence of team members.
☐	☐	☐	☐	19. Our team leader makes sure that team members are clear about critical issues and important facts.
☐	☐	☐	☐	20. Our team leader exhibits trust by giving us meaningful levels of responsibility.
☐	☐	☐	☐	21. Our team leader is fair and impartial toward all team members.
☐	☐	☐	☐	22. Our team leader is an optimistic person who focuses on opportunities.
☐	☐	☐	☐	23. Our team leader looks for and acknowledges contributions by team members.

IV.   Demonstrate Sufficient Technical Know-How

True	More True Than False	More False Than True	False	
☐	☐	☐	☐	24. Our team leader understands the technical issues we must face in achieving our goals.
☐	☐	☐	☐	25. Our team leader has had sufficient experience with the technical aspects of our team's goal.
☐	☐	☐	☐	26. Our team leader is open to technical advice from team members who are more knowledgeable.
☐	☐	☐	☐	27. Our team leader is capable of helping the team analyze complex issues related to our goal.
☐	☐	☐	☐	28. Our team leader is seen as credible and knowledgeable by people outside our team.

VII.   Set Priorities

True	More True Than False	More False Than True	False	
☐	☐	☐	☐	29. Our team leader keeps our team focused on a manageable set of priorities that will lead to the accomplishment of our goal.
☐	☐	☐	☐	30. Our team leader and the members of our team agree on the top priorities for achieving our goal.
☐	☐	☐	☐	31. Our team leader communicates and reinforces a focus on priorities.
☐	☐	☐	☐	32. Our team leader does not dilute the team's effort with too many priorities.
☐	☐	☐	☐	33. If it's necessary to change priorities our team leader helps us understand why.

VIII.   **Manage Performance**

True	More True Than False	More False Than True	False	
☐	☐	☐	☐	34. Our team leader makes performance expectations clear.
☐	☐	☐	☐	35. Our team leader encourages the team to agree upon a set of values that guides our performance.
☐	☐	☐	☐	36. Our team leader ensures that rewards and incentives are aligned with achieving our team's goal.
☐	☐	☐	☐	37. Our team leader assesses the collaborative skills of team members as well as the results they achieve.
☐	☐	☐	☐	38. Our team leader gives useful, developmental feedback to team members.
☐	☐	☐	☐	39. Our team leader is willing to confront and resolve issues associated with inadequate performance by team members.
☐	☐	☐	☐	40. Our team leader recognizes and rewards superior performance.

41.   What are the strengths of the team leader?

42.   What one or two changes are most likely to improve the effectiveness of the team leader?

APPENDIX C

Manager Performance Rating Survey

I.   Rating of Team Performance

Rate the performance of each one of your assigned teams with a score of 4.0 (highest) to 1.0 (lowest)

<u>Team Type
& Number</u>   <u>Score</u>

1. _____   _____

2. _____   _____

3. _____   _____

4. _____   _____

5. _____   _____

6. _____   _____

7. _____   _____

8. _____   _____

9. _____   _____

10. _____   _____

11. _____   _____

12. _____   _____

13. _____   _____

14. _____   _____

15. _____   _____

II.   **The three most important criteria for an effective team are:**

1.

2.

3.

APPENDIX D

Project Approvals

-----Original Message-----
From: George DeMetropolis
Sent: Thursday, October 10, 2002 8:41 AM
To: HR Company Representatives
Cc: Project Management
Subject: Survey of Team Personnel

I have discussed my intent and received authorization to conduct a survey of team personnel from upper project management. The focus of the survey is to identify what actions and demeanor a team leader uses to create an effective and productive team. It is my hope that I will be able to identify these relationships, so that we can further hone the skills of those teams that are less effective to become more productive contributors for this project. I will also use this information to fulfill the partial requirements for a dissertation.

Participation in this survey is on a voluntary basis and personnel may refuse to participate at any time without any consequence or prejudice. The survey involves filling out a survey instrument, which is attached below for your information. No individual will be identified in any way. No one will see the completed forms other than an Area Supervisor or Senior UXO Specialist Supervisor/UXO Demolition Manager and me. None of these personnel nor I will discuss individual participants with anyone. The data provided will be recorded anonymously and participation and anything disclosed will be held in the strictest confidence. Both confidentiality and privacy will be protected.

I will share the results of this survey with you as soon as summary information is available. Thanks for your support.

----Original Message-----
From: Union Representative
Sent: Monday, October 21, 2002 3:32 AM
To: George DeMetropolis (E-mail)
Subject: Evaluations

George,

One of the company representatives contacted me regarding
"Evaluations" of personnel on the project. Would you please bring me up
to speed on what it is that you need? Is there anything that you want from
me?

Mark

-----Original Message-----
From: George DeMetropolis Sent: Monday, October 21, 2002 9:18 AM
To: Union Representative
Subject: RE: Evaluations

Hi Mark,
Good to hear from you. I am fine and continue to enjoy this terrific project
because of the outstanding people that we work with. Attached is a
message that describes what I am attempting to do and accomplish. The
focus of the survey is the UXO team leader. Members of the team will
have the opportunity to comment on how they perceive the team to be
doing.

Please let me know if there is any other information needed for you to
allow me to proceed. Thanks, Mark.

<<Survey of Team Personnel>>

181

-----Original Message-----
From: Union Representative
Sent: Monday, October 21, 2002 6:41 AM
To: 'George DeMetropolis'
Subject: RE: Evaluations

George,

Thanks for the information. It appears that the information that you are
trying to obtain will come, in part, from represented workers.
Personally, I think it is a good idea to identify those characteristics so that
we can build upon that. However, I do want to pass this on to the Union
folks that I deal with to get their buy-in. Since it voluntary, I do not see
where there will be any resistance.

Mark

-----Original Message-----
From: George DeMetropolis
Sent: Monday, October 21, 2002 9:59 AM
To: Union Representative
Subject: RE: Evaluations

Thanks for the quick response, Mark. I will hold off until I hear from you.

-----Original Message-----
From: Union Representative
Sent: Wednesday, October 23, 2002 4:10 AM
To: 'George DeMetropolis'
Subject: RE: Evaluations

George,

I just got word from the Union that they do not have a problem with it as
long as it is voluntary.

Good luck! Hope to see you soon.
Mark

APPENDIX E

Administrator's Briefing Sheet

- ➢ We have discussed our intent and received authorization to conduct an island wide survey of team personnel from project, union, and company representatives.

- ➢ The focus of the survey is to identify what actions and demeanor a team leader uses to create an effective and productive team. It is our hope that we will be able to identify these relationships, so that we can further hone the skills of those teams that are less effective to become more productive contributors for this project.

- ➢ Participation in this survey is on a voluntary basis and personnel may refuse to participate at any time without any consequence or prejudice to include promotions or further employment.

- ➢ The survey involves filling out a survey questionnaire. No individual will be identified in any way. No one will see the completed forms other than an Area Supervisor or Senior UXO Specialist Supervisor/UXO Demolition Manager and the Range Control Officer. None of these personnel nor I will discuss individual participants with anyone. The data provided will be recorded anonymously and participation and anything disclosed will be held in the strictest confidence. Both confidentiality and privacy will be protected.

- ➢ Results of this survey will be discussed with you as soon as summary information is available. Thanks for your support.